KARL MARX AND MATHEMATICS

A Collection of Texts in Three Parts

Edited by

Pradip Baksi

AAKAR

Karl Marx and Mathematics: A Collection of Texts in Three Parts
Edited by Pradip Baksi

ISBN 978-93-5002-595-6

Data for Cataloguing the Publication:
(1) The history of, (2) investigations on and, (3) plural mathematics before and after, the mathematical manuscripts of Karl Marx.

Published by
AAKAR BOOKS
28 E Pocket IV, Mayur Vihar Phase I
Delhi 110 091, India
www.aakarbooks.com

Designed by
Limited Colors, Delhi

Printed at
D.K. Fine Art Press, Delhi

SIGN AND ABBREVIATIONS

Words, clauses or sentences within <   > brackets indicate subsequent insertion for the present volume.

MEGA: Marx-Engels-Gesamtausgabe (1975-).
Berlin: Dietz-Verlag / Akademie Verlag / De Gruyter Akademie Forschung.

MECW: Karl Marx / Friedrich Engels, *Collected Works* (1975-2005), Moscow: Progress etc.

Contents

Part Two: Investigations

Part Three: Plural Mathematics

Introduction, 2019

This collection of various texts on Karl Marx and Mathematics is the revised and extended second edition of the Special Supplement to Karl Marx, *Mathematical Manuscripts* (1994; Calcutta: Viswakos) titled Marx and Mathematics.

A part of the additional information included in the present collection has evolved during the last 25 years. Some of the relevant developments of this period have been indicated in the Translator's Note to the revised second edition of Karl Marx, *Mathematical Manuscripts* (2018: xxxiv-xxxvi; Delhi: Aakar Books). The remaining part of the information indicated in this introduction and collection existed but was either inaccessible to most people before the emergence and development of the World Wide Web or, was neglected by the earlier editors, including the present author. The sources of the texts included in the three parts of this collection and, some biographical information about their respective authors have been indicated at the end of each text.

Today, digital copies of the source texts studied by Marx, and those of his mathematical notes, excerpts and manuscripts are available in the public domain. Though currently postponed *sine die*, these manuscripts, notes and excerpts remain slated to be published within the *Marx-Engels-Gesamtausgabe* (*MEGA*[2]) as *MEGA*[2] I/28 and IV/30. Studies in the history of Marx/Engels editing and publishing, and of its role in the emergence and development of the ideologies like Marxisms, Leninism, and Stalinism etc. continue to register progress [Leckey, 1995; Marxhausen, 2006; Mosolov, 2010; Sperl, 2005 and 2016].

The emergence and development of the Ethnomathematics movement continue to change our understanding of the history of evolution of plural mathematics on planet earth since the Neolithic age. Rediscovery and study of some of the neglected source texts have further energized investigations on the subsequent history of mathematical cultures, including those on the histories of algebra and analysis in some of the ancient and medieval languages of Asia, like Sanskrit, Arabic and Malayalam. Consequently, it is now possible to indicate some of the larger gaps in the dominant understanding of history of mathematics not only in Marx's time, but also at the time of editing Marx's mathematical manuscripts in the twentieth century, and even today.

Finally, the emergence and development of mathematical and statistical software packages are vigorously reshaping our ways of conceptualizing and doing mathematics towards an unknown future. It is time now for taking yet another look at all mathematical text from the past and that includes the mathematical manuscripts of Marx.

The present introduction has three parts, pertaining to some aspects of: 1) the history of editing and reception of Marx's mathematical manuscripts; 2) the past history and present developments of mathematics within which the said manuscripts were and are being edited and studied; and, 3) some of the present and possible future issues of interest to the students of Marx and mathematics.

EDITING AND RECEPTION

The first editor of Marx's mathematical manuscripts during the years 1925-27 and 1930, Emil Julius Gumbel [1891-1966] is known as the principal architect of the statistical theory of extreme values, as a fearless crusader against political murder by the Nazis in Germany, and as a German-American statistician and pacifist:

<https://www.encyclopediaofmath.org/images/a/ad/EmilJuliusGUMBEL.pdf>

For a narrative on this first period of editing of the mathematical manuscripts of Marx, see: Vogt, 1995.

In a communication published in 1927, Gumbel noted that Marx had mainly studied some of the mathematical texts known towards the end of the eighteenth century and, that his choice of texts was influenced by that of Hegel. Hegel mentioned the same works of Euler, Lagrange, Landen and Newton that Marx used or mentioned and, that like Hegel, he too was preoccupied with the foundations of differential calculus, and with the process of deduction of derivatives. He pointed out further, that Marx had no direct acquaintance with preceding or contemporary mathematical literature, such as the works of Carl Friedrich Gauss [1777-1855] or those of Bernard Bolzano [1781-1848] (Gumbel, 1927: 59).

Gumbel's statement on the interrelationships of Hegel and Marx in the domain of mathematics needs reexamination in the light of subsequently published literature; for a bibliography on Hegel and Mathematics see: Appendix I.

Gumbel was entrusted with the task of editing Marx's mathematical manuscripts by David Borisovich Riazanov [1870-1938], the founder-director of the Institute of K. Marx and F. Engels [IME, 1921-1931] in Moscow:

<http://rjazanov.shpl.ru/>

Towards the end of Stalin's campaign against his adversaries within the leadership of the ruling party of USSR and, at the very beginning of the emergence of his cult in 1931, Riazanov was removed from his job at that institute, expelled from the party, and arrested. He was shot dead at Saratov in 1938, formally rehabilitated in 1958, and politically rehabilitated in 1989.

Sometime after Riazanov's arrest in 1931, the editorial responsibilities related to Marx's mathematical manuscripts were handed over to Sofya Aleksandrovna

Yanovskaya [1896-1966], a member of the Bolshevik party since 1918, a former commissar of the Red Army during the Civil War, a Professor of the Moscow State University, the Institute of Red Professors and the Academy of Sciences of the USSR since 1931 and, one of the principal architects of the study of philosophy and history of mathematics and, mathematical logic in the USSR:

<http://www-groups.dcs.st-and.ac.uk/history/Biographies/Janovskaja.html>

<http://www-groups.dcs.st-and.ac.uk/history/References/Janovskaja.html>

At that time the Institute of Marx and Engels [IME] had already become the Institute of Marx, Engels and Lenin [IMEL, 1931-1954], on its way to becoming the Institute of Marx, Engels, Lenin and Stalin [IMELS, 1954-1956]; and finally, the Institute of Marxism-Leninism [IML, 1956-1991].These changes reflect the shifting ideological stances and priorities of the rulers of USSR, over the last 60 years of the existence of that state.

A year before becoming the second editor of Marx's mathematical manuscripts, Yanovskaya wrote that there were very few "mathematician-marxists" in the USSR; that the students working to obtain doctorate degrees in mathematics—if not loudly, then at least noiselessly—called the courses on history and philosophy of the natural sciences that were organized for them to be "red theology"; and, that the old professors of the "Moscow School" of mathematics spared no efforts to save the autonomy of "pure" mathematics from the malicious assaults of the partisans of materialist philosophy. Even the word "Comrade" was not permitted in the Institute of Mathematics and Mechanics, within the Mathematical Society and, in their journal the *Matematicheskii sbornik* [*Mathematical Collections*], as quite a significant percentage of its members were white émigrés (Yanovskaya, 1930: 88). She further reported that at that time in the Mathematics Department of the Faculty I of Physics and Mechanics of the Moscow State University there were only 2 full members and 1 candidate member of the ruling party and, only 3 members of the Komsomol, all about to pass out (ibid.: 90). Throughout this article she lamented the lack of, and offered her proposals for, nurturing mathematically well-trained and ideologically well-fortified marxist-leninist cadres for the "mathematical front". To trace the evolution of some of her ideas as a party ideologue embedded within the academic bureaucracy of mathematics in the USSR in the 1930s, see: Yanovskaya, 1931 and 1936. For some overviews of the history of interrelationships of mathematical theories, applied mathematics, religion, ideology, philosophy and politics in Russia and the Soviet Union, see: Vucinich, 1999, 2000 and 2002; Lorentz, 2002; Seneta, 2004; Graham and Kantor, 2009; Hollings, 2013; and, Demidov and Lëvshin, 2016.

In 1932, Yanovskaya studied the mathematical manuscripts of Karl Marx in collaboration with Dmitri Rykov of Moscow and A.E. Nahimovskaya from Minsk (Rosenfeld, 1996: 74). In a long footnote attached to an article published in 1933 Yanovskaya acknowledged that the "photocopies of Marx's mathematical manuscripts were available at the Institute of Marx-Engels since 1925" and, that "a significant part of the said manuscripts were quite scrupulously deciphered by 1927" (Yanovskaya, 1933: 75-76, footnote 2). However, she noted that "the materials chosen for deciphering and their presentation were not that satisfactory" (ibid.).

This was, above all, the case with Marx's most valuable work, the "Historical Essay", which provided a key to the rest of the texts. This work was listed among "excerpts from various authors" as there were changes in the modes of notation (the reference is to Gumbel, 1927: 58; translation in Marx, 2018: xi). However, Yanovskaya continued, "it contains no excerpts and, the modes of notation have changed because it contained discussion about the different stages of development of differentials calculus" (ibid.). At the same time, some explicit notes (related to the divergent serieses and to the Taylor's theorem) were placed within the slot of "independent works". She continued: "The presentation of the works on the theorems of Taylor and MacLaurin were especially disorderly; there, often one page from one work was followed by another page from another work"(ibid.). Yanovskaya opined that all of this is fully explained by the "careless attitude to work" on the part of the previous leadership of the Institute (meaning thereby Riazanov and his colleagues) and, by a "complete non-understanding" of those mathematical manuscripts of Marx that have already been put in order. To "prove" that such was indeed the case, Yanovskaya provided the following quotation from Gumbel, 1927: 59: "Those manuscripts of Marx that do not contain calculations or excerpts may be considered to be his independent work. Publishing only those makes sense. They contain *a free exposition of what Marx studied*, combined with numerous *chronological dates, and philosophical reflections on what he read*; in some cases Marx posed some problems in a special way that is characteristic only of him" (Yanovskaya's stress).Her article contains two laudatory ideological references to some philosophical statements by Stalin (on pages 102 and 110), without indicating their textual sources.

Subsequently some of the ideological standpoints of Sofya Aleksandrovna Yanovskaya about mathematics, philosophy and politics changed. Her intellectual evolution through self-education, reflections on her own political activities, experience over the years and, through her interactions with other people has been traced by some authors (Bazhanov, 2001; Levin, 2017). There exists an opinion that her orientations on mathematics, logic and philosophy underwent gradual but significant changes after her meetings with Ludwig Josef Johann Wittgenstein [1889-1951] at her university and her flat, during Wittgenstein's visit to Moscow, in September 1935 (Biryukov and Biryukova, 2004; Vasiliev, 2017a and 2017b). Wittgenstein also met the mathematician and logician Valery Ivanovich Glivenko [1897-1940] and, the historian of mathematics Adolf-Andrei Pavlovich Yushkevich [1903-1996] during that trip to Moscow. Prior to this meeting Yanovskaya was primarily concerned with history and philosophy of mathematics and worked on the mathematical manuscripts of Marx. Since 1936 she concentrated more and more on the study, translation and teaching of mathematical logic; on the evolution of this aspect of her work see: Küng, 1962; Anellis, 1996; Bazhanov, 2001. During World War II in the USSR [1941-1945], Sofya Aleksandrovna Yanovskaya was evacuated to Molotov/Perm. There she taught algebra at the Molotov/Perm State University (Rosenfeld, 1996: 75).

Since the arrest, removal, and finally murder of David Borisovich Riazanov, the work related to editing and publishing the works of Marx and Engels lost its earlier momentum. Many factors contributed to this decline. At the time of removal of Riazanov from the IME some 131 of its 243 employees were also removed (Rokityansky, 2008: 20), creating a severe shortage of the required editorial staff. The ruling party of the USSR was more interested in publishing the works of Lenin and Stalin in the interests of shaping and promoting the ideologies of Leninism and Stalinism. Not all texts of Marx were found to be equally useful for manufacturing the ideology of Marxism, subsequently combined with Leninism and, rebranded as Marxism-Leninism. 10 years after the removal of Riazanov the publication of the *Marx-Engels-Gesamtausgabe* [*MEGA*^{1}] was stopped. After the publication of some of Marx's texts related to the derivative, differential and history of the differential calculus in the journal *Pod znamenem marksizma*, in 1933 (No. 1: 15-74), only his notes on the concept of function was published in the *Voprosy filosofii* in 1958 (No. 11: 89-95). The details of the work done and/or not done in respect of editing the mathematical manuscripts of Marx during the years 1933-1966 may become explicitly clear, only if the necessary archival materials exist and, if those records are investigated by the concerned scholars in future. The same holds good for the gap years in the history of the *MEGA,* which resumed its second journey as *MEGA*^{2} in the year 1972.

Towards the end of her life Yanovskaya appears to have endorsed, and even extended, some of the observations of Gumbel on Marx's mathematical manuscripts (Yanovskaya, 1968). She and some of the subsequent translators and editors of these manuscripts in other languages, and those who reviewed those publications, took note of the fact that Marx had no direct acquaintance with the relevant works of Augustin-Louis Cauchy [1789-1857], Karl Theodor Wilhelm Weierstrass [1815-1897], Richard Dedekind [1831-1916] and Georg Ferdinand Ludwig Philipp Cantor [1845-1918]. Those among these editors who were exposed only to this history and practice of what is now called the classical mathematical analysis opined that consequently Marx's efforts to grapple with the foundations of differential calculus remained somewhat dated even for his own time.

Yanovskaya was exposed to the history of intuitionist and constructive mathematical analyses starting from the works of Luitzen Egbertus Jan Brouwer [1881-1966], right through those of Andrei Andreevich Markov [1903-1979], Arend Heyting [1898-1980], Errett Bishop [1928-1983], Andrei Nikolaevich Kolmogorov [1903-1987] and their followers (on this history see: Nepeyvoda, 1987). She observed: "It is true that Marx was not acquainted with the modern definitions of the concepts of real number, limit and continuity. However, it seems that even if he had an acquaintance with these definitions, he would not have been satisfied with them. The situation is as follows: Marx was in search of the "real" process of finding out the derived function, i.e., of an algorithm, which will permit, firstly, to answer the question whether there exists a derivative for a given function, and secondly—if it exists then how to find it out effectively. It is well known that the concept of limit is not algorithmic, and that is why such problems are solvable only for a certain class of functions. One of those classes is the class of analytical functions, i.e., the class of such functions, which are decomposable into powered serieses and—to use an expression of Marx—are objects of "algebraic" differentiation. As a matter of fact only such functions were investigated by Marx. At present the class of such functions—for which the two aforementioned questions can be answered—may be significantly extended, and operations with them may be so constructed, as to satisfy all the modern demands of strictness and exactitude. However, from Marx's point of view it is essential, that all passages to limit be investigated in the light of their effective completion; in other words, it is essential to construct mathematical analysis basing it upon the theory of algorithms, as we would put it now" (Predislovye, Marks, 1968: 11-12; Preface, 1968 in: Marx, 2018: xxii).

Sofya Aleksandrovna Yanovskaya died in 1966, two years before the 1968 Moscow edition of Marx's mathematical manuscripts came out in print. This edition is silent about the editorial work done by Emil Julius Gumbel, and the name of Yanovskaya has been mentioned only once. The emergence and development of nonstandard methods of analysis from around the same time, especially since the publication Abraham Robinson's seminal work in 1966, led to a different evaluation of the history of differential calculus and, consequently, also of Marx's mathematical manuscripts.

There exists a remark of Kurt Gödel [1906-1978] within the Preface to the second edition of the *Non-standard Analysis* by Abraham Robinson (1974). Gödel said: "Arithmetic starts with the integers and proceeds by successively enlarging the number system by rational and negative numbers, irrational numbers, etc. But the next quite natural step after the reals, namely the introduction of infinitesimals, has simply been omitted. I think in coming centuries it will be considered a great oddity in the history of mathematics that the first exact theory of infinitesimals was developed 300 years after the invention of the differential calculus" (Gödel, 1990: 311). Gödel expressed the hope that in some version or the other, nonstandard analysis containing the exact theory of infinitesimals, "will be the analysis of the future" (ibid).

Within a couple of years Howard Jerome Keisler reintroduced infinitesimals into the classroom (1976). Two Chinese translations of Marx's mathematical manuscripts from their German, Russian and Japanese renderings came out in Shanghai and Beijing in 1975. There these translations and their study got intertwined with the emerging study of nonstandard analysis and a raging ideological battle around Maoism (Dauben, 1998 and 2004; Xu, 2013). In the same decade Hubert Kennedy presented a paper on Marx, Peano, and the Differentials at the 15th International Congress of History of Science, at Edinburgh (Kennedy, 1977). For some accounts of the Soviet receptions of Marx's mathematical manuscripts, see: Molodshii, 1983; Przhesmitsky, 1985; Ruzavin, 1987. Adolf-Andrei Pavlovich Yushkevich wished to call all mathematics emerging around that time to be "non-standard mathematics" (1987); and, Feodor Andreevich Medvedev [1923-1993] attempted to situate nonstandard analysis in the context of history of classical analysis (1987). For some glimpses into the state of philosophy of mathematics in the USSR around that time see: Uspensky, 1987.

Perhaps, towards the end of 1987 a decision was taken to prepare a fresh and complete edition of Marx's mathematical manuscripts for the *MEGA*².This time the task of editing these manuscripts was entrusted to Irina Konstantinovna Antonova of the Institute of Marxism-Leninism of CC CPSU, at Moscow. Her work was slated for review upon completion, by some other experts of GDR and USSR belonging to the editorial board of the *MEGA*². In 1990 the newly formed Internationale Marx-Engels-Stiftung (IMES) [International Marx-Engels Foundation], established at the initiative of the Internationaal Instituut voor Sociale Geschiedenis (IISG) [International Institute of Social History (IISH)] of Amsterdam, assumed the responsibility of publishing the *MEGA*², as the earlier sponsors of this project, the governments and state structures of GDR and USSR collapsed. A process of reorganization followed. The new management decided to publish the mathematical manuscripts of Marx as *MEGA*² I/28 and IV/30.

The new editorial team for *MEGA*² I/28 consisted of Irina Konstantinovna Antonova of the newly constituted Russian State Archive of Social and Political History together with, Alain Alcouffe and Jean-Claude Yakoubsohn of the Université de Toulouse I, France. These names are still indicated at: <http://www.iisg.nl/imes/mega1.php>. The editorial team for *MEGA*² IV/30 was not constituted. In May-June 1995, the material conditions of the people of Russia, including those of the employees of the erstwhile Institute of Marxism-Leninism deteriorated very sharply: the corridors of the "Institute" were cluttered with bags of sugar received from Cuba, which the employees were forced to trade for food. Irina Antonova had to work as a guide for tourists in order to survive. To get a feel of the problems faced by the editors of the MEGA around that time see: Kurbjuweit, 1993. Jürgen Rojahn (of the IMES and IISG), Hans Pelger (of the Karl-Marx-Haus, Trier) and Jacques Grandjonc (the head of the Editorial Commission of IMES) did their best to find some funding for the Moscow team which was in charge of the Mathematical Manuscripts of Marx and of some other volumes of the *MEGA*² through various European Union programs. They were concerned with the completion of the work in progress. In this connection Alain Alcouffe was in Moscow from the 12th to the 19th of October 1995, to discuss with Georgii Aleksandrovich Bagaturia and Irina Konstantinovna Antonova about how to complete the volume which should include the *Mathematical Manuscripts* of Marx. Irina Konstantinovna had almost finished typing the manuscripts using the contemporaneous version of Microsoft Word; but she was soon obliged to leave the newly constituted Russian State Archive of Social and Political History. After a week of discussion a zip drive containing the files was given to Alain Alcouffe; he also received a complete set of hard copy of the *Mathematical Manuscripts*. The zip drive was also sent to the Berlin-Brandenburgische Akademie der Wissenschaften (BBAW) where a specialist of Marx's handwriting was supposed to check the dates assigned to the texts in course of redaction/editing of the said manuscripts.

Unfortunately, the respective administrative responsibilities were not clearly established, let alone the financial arrangements. Editorial responsibilities seemed fairly easy to divide: the production of the manuscript was entrusted to the BBAW, for verification of conformity with Marx's handwritten texts and for dating of the various manuscripts. Alcouffe and Yakoubsohn were to compare the document produced by the BBAW with the sources used by Marx that were largely French textbooks. A researcher of the BBAW, Carl-Erich Vollgraf was to go to Amsterdam to examine the originals at the IISG. C.-E. Vollgraf was unable to travel to Amsterdam and compare the output of Irina Konstantinovna Antonova with the originals before 2016. After spending some time at the IISG, he concluded that the distribution of the manuscripts between I/28 and IV/30 did not have a strong rationale. Eventually the publication of the mathematical manuscripts of Marx was postponed *sine die*. Now no editorial teams are indicated within the web portal of the BBAW, for these two volumes of the *MEGA*². Information about how long the members of the above indicated team worked, under what circumstances and, what results they produced is not available in the public domain. 50 years after the publication of the 1968 Moscow edition, the future of Marx's mathematical manuscripts seem once again to be uncertain, this time within the current German academic dispensation of the *MEGA*². The reconstruction of this period from 1968 to 2019, plus that of the earlier indicated period from 1925 to 1966, that is, of the last 94 years of the history of editing of the mathematical manuscripts of Marx, from the institutional and personal archival materials available at Moscow, Amsterdam, Berlin and Toulouse, remains a part of the task for the future historians these manuscripts.

One member of the last editorial team, Irina Konstantinovna Antonova attempted a critical analysis of Marx's mathematical manuscripts in an article; and, a reconstruction of the process of cognition on the basis of the same manuscripts in a book chapter (Antonova, 1999 and 2004). Another member, Alain Alcouffe, editor of the French translation of Marx's mathematical manuscripts (1985), published the results of his investigations on Marx's study of differential calculus from a textbook by Jean-Louis Boucharlat [1775-1848], in the light of the study of the same text by Paul Matthieu Hermann Laurent [1841-1908] and, that of further developments in the history of analysis (Alcouffe, 1987). Subsequently, in a working paper on Marx, mathematics and the *MEGA*², Alain Alcouffe and Julian Wells highlighted Marx's interest in statistical and probabilistic ideas (Alcouffe and Wells, 2009).

In the wake of contemporary developments in nonstandard methods of analysis and Leibniz studies we learn that the "mysterious mist" surrounding the earlier conceptions of the infinitesimals does not so much belong to the pioneers of differential calculus as to their followers. It was not the creation of a deep thinking Leibniz. It was created by the over simplifying Leibnizians. Consequently, a more careful reading of Marx's texts informed by an appreciation of nonstandard analysis is in order (Kusraev and Kutateladze, 1994: 1-12, especially pages 6-8 therein). More recently, it has also been proposed in the light of some earlier French scholarship (Alcouffe, 1985: 9-108) that Marx's mathematical manuscripts need to be viewed not as a finished treatise, but rather as active reflections on doing and learning differential calculus and, that Marx's reflections drew their inspirations from the mathematical reflections of Hegel who too aspired for a secure foundation for the infinitesimals; such security was provided at long last by the nonstandard analysis (Serman, 2013).

For a bibliography on Hegel and mathematics, see: Appendix I. Augusto Ponzio, the famous semiologist, philosopher and, editor of the Italian translation of Marx's mathematical manuscripts (2005) has placed these manuscripts in the context of the semiotic practices of Karl Marx (Ponzio, 2014). For a bibliography on mathematics and semiotics, see: Appendix II.

Thanks to an indication in a list published in *MEGA* IV/3 (1998: page 7, line 29) we also came to know about the existence of a possible mediating link between Hegel and Marx in mathematics: it is reportedly the first Hegelian treatment of philosophy of mathematics, by Constantin Frantz [1817-1891] (Frantz, 1842). Further, a text indicating Marx's use of mathematics within the manuscripts of what came to be known as his "*Capital",* Volume III: "Mehrwertrate und Profitrate mathematisch behandelt" ["Mathematical treatment of rate of surplus value and rate of profit"] and Samuel Moore's summary of and observations on them contained in a letter to Engels dated 23 March 1888, have also been published as parts of the *MEGA* II/14 (Marx and Engels, 2003: 19-150 and 351-359).

That continuously fresh reading of Marx's mathematical manuscripts is both necessary and possible was also indicated in part XII of Marx's incomplete manuscript titled "Taylor's Theorem". There Marx first obtained the derivative of x^m for integral and positive m using the formula for the differential of product. Then he demonstrated this formula by the methods of Newton-Leibniz, evidently thinking that since we have already operated upon the grounds of differential calculus, in other words, since we have substantiated its methods, we may now proceed to use the methods of Newton and Leibniz too: that is, we may consider them sufficiently substantiated (Marks, 1968: 557-558; Marx, 2018: 307). Yanovskaya commented in this context: "Thus, it will be extremely unfair to depict the standpoint of Marx as the rejection of all other methods used in differential calculus. If these methods turn out to be successful, then Marx sets before himself the task of explaining the secret of their success. When he succeeds in this, i.e., after the method under consideration turns out to be well grounded and the conditions of its applicability are fulfilled, then Marx considers a transition to that method not only entirely justified, but also expedient" (Predislovye, Marks, 1968: 15; Preface, Marx, 2018: xxv).

We have seen so far that the histories of editing and reception of Karl Marx's mathematical manuscripts have unfolded at the intersection of two larger histories: (1) that of mathematical analysis and, (2) that of editing, publishing and reception of the works of Marx and Engels before and after the collapse of USSR and GDR.

For contemporary narratives on the history and historiography of mathematical analysis, see: Sonar, 2011 and Sinkevich, 2014. It appears today that in the realm of differentiation of analytical functions Marx may be considered to be one of the neglected precursors of both the constructivist, and the nonstandard trends. The various earlier and currently emerging trends in the nearly three thousand years old history of analysis may be viewed as outcomes of a continuing search for exactitude without surrendering inventiveness. It has been indicated by the earlier editors, commentators and critics that Marx's study of this process, through his investigations into the nature and history of differentiation of analytical functions, was handicapped by information gaps, by his lack of acquaintance with the texts reflecting the relevant developments of the nineteenth century. It has largely gone unnoticed, however, that all the major mathematicians tackled by Marx, all of the textbooks that he used, the single text on history of mathematics that he consulted and, finally all the mathematicians of the twentieth century who edited his manuscripts were also, without exception, handicapped by major information gaps and, were influenced by the dominant paradigms of the mathematical cultures of their geographical space and historical time.

PAST AND PRESENT

Karl Marx studied algebra, commercial arithmetic, analytical geometry, trigonometry and differential calculus from the early 1850s to the early 1880s; see: Karl Marx/ Friedrich Engels Papers, A 100-104 and a part of B 142 located at:

<https://search.socialhistory.org/Record/ARCH00860>.

The authors of the texts used by Marx and the editors of Marx's mathematical manuscripts were either unaware of or they just ignored some of the published relevant information about the nature and history of these disciplines in different cultures. This information was published in some of the very high profile journals of the time, which were certainly available in London, and hopefully also in Paris, and Berlin, during the 18th-19th centuries. Here are some examples.

1790 and 1792: The papers of Reuben Burrow [1747-1792] on the Binomial Theorem of the Hindoos (1790) and, on one of the Hindoo Rules of Arithmetic (1792), in the *Asiatick Researches,* Volumes II and III. In the first paper Burrow had expressed the view that the ancient Indians "were acquainted with a *differential method similar to Newton's*" (Burrow, 1790: 491; my stress). The second paper is about an ancient Indian Rule for *"finding the sum of all the different permutations of a given numeral quantity, consisting of a given number of places of figures"* (Burrow, 1792: 145; author's stress).

Pramatha Nath Bose [1855-1934] observed in the *Centenary Review of the Asiatic Society of Bengal 1784 to 1883*: "The earliest notices of Indian Algebra and Arithmetic which reached Europe were from Burrow. He made a good collection of mathematical manuscripts, which he bequeathed to his friend, Mr. <Isaac>Dalby<1744-1824>, of the Royal Military College, who communicated them to those, interested in the subject in 1800" (Bose, 1885, III.I, §8: 22).

Reuben Burrow was one of the friends and teachers of Allama Tafazzul Hussain Khan [1727-1800], a bright star in the firmament of what came to be known as the Awadh scientific renaissance (Khan, 2003). He translated Newton's *Principia,* Apollonius' *De Sectione Rationis,* another treatise on conic sections by Guillaume François Antoine Marquis de l'Hôpital [1661-1704], a treatise on algebra prepared by Burrow for him and, the *Mechanics* of William Emerson [1701-1782] into Arabic during his stay at Calcutta, as the ambassador of Awadh (*The Asiatic Annual Register for 1803,* London: 1804; the section titled Characters: 7).

1817: The translation by Thomas Henry Colebrooke [1765-1837] of the algebra, arithmetic and mensuration of Brahmagupta [c.598-c.668] and Bhāskarāchārya [1114-1185] published in London (Colebrooke, 1817).

1820, 1832 and 1837: The essays of John Tytler [1790-1837] on the Binomial Theorem as known to the Arabs (1820); on the extraction of Roots of Integers as practised by the Arabs (1832); and, an Analysis and Specimens of a Persian Work on Mathematics and Astronomy (1837).

Read in 1832 and published in 1834: The report of Charles Matthew Whish [1794-1833] on the Hindu Quadrature of the Circle, and the infinite series of the proportion of the circumference to the diameter exhibited in four texts in Sanskrit and Malayalam dating back to the 14th-16th centuries (Whish, 1834).

1837: The notice of Lancelot Wilkinson (1805-1841) on the discovery of the manuscripts of the *Rekhā-gaṇita,* a translation into Sanskrit by Paṇḍita Jagannātha Samrāṭ [c.1657- c. 1744] of the Arabic recension of the *Elements* of Euclid [c. 300 B. C.] by Nasir al-Din al-Tusi [1201-1274] (Wilkinson, 1837).

1850 and 1859: Editor's Preface by Augustus De Morgan [1806-1871] to *A Treatise on Problems of Maxima and Minima, solved by Algebra,* by Master Ramchundra [1821-1880] of Delhi College (Ramchundra, 1850; Reprint 1859: iii-xxiii).

1858:An article by Bapu Deva Shastri [1821-1900] on Bhāskarāchārya's knowledge of the differential calculus (Shastri, 1858).

1860: The critical note of William Spottiswoode [1825-1883] on the article of Bapu Deva Shastri indicated in Shastri 1858 above (Spottiswoode, 1860).

1865: Brief notes by Bhāu Dājī [1822-1874] on the age and authenticity of the Works of Āryabhaṭa, Varāhamihira, Brahmagupta, Bhaṭṭotpala, and Bhāskarāchārya (Dājī, 1865).

For some of the subsequent discussions related to mathematics in ancient and medieval South Asia, see: Sengupta, 1932; Datta and Singh, 1935, 1938, 1962; Datta, Singh and Shukla, 1984; Bag, 1966 and 1979; Weil, 1984; Pingree, 1970-1994; Gupta, 2002; Keller, 2006; Sarma, 2008; Divakaran, 2012; and, Smorynski, 2017. For some of the subsequent studies related to the history of mathematics in Arabic see: Rashed, 1994 and 2015.

How the internalization of the information indicated in the vast literature mentioned above would have influenced the work of the pioneering mathematicians and influential textbook writers of Europe in the fields of algebra and differential calculus and, also the corresponding discourse of history of mathematics is anybody's guess. Marx's lack of exposure to the frontiers of mathematical analysis of his time requires to be viewed in the context of this gigantic information gap among those mathematicians of Europe, who came before and after Marx's time. There is no point in lamenting what they did not know, learn and, consequently what they failed to transmit to the self-educated students of mathematics like Karl Marx. Some historians of mathematics (Scriba and Schreiber, 2001; Alten *et al,* 2003; Wußing, 2008 and 2009; Roy, 2011) and textbook writers (Körle, 2015) have taken some steps to integrate the hitherto unutilised relevant information into their respective narratives.

For mapping of the corresponding information gaps also in respect of mathematics in ancient and medieval China, Mesopotamia, Egypt, Europe and North Africa one may start with: Martzloff, 1987; Clagett, 1999; Dold-Samplonius *et al,* 2002; Katz *et al*, 2007 and 2016; and, Imhausen, 2016.

Some of the online resources available for the present and future students of Karl Marx and mathematics include:

Catalog of Mathematics Resources on the Web:
<http://www.uwc.edu/depts/math/resources/catalog>;

MacTutor History of Mathematics Archive:
<http://www-groups.dcs.st-andrews.ac.uk/~history/>;

Selin, Helaine (2000), *Mathematics across cultures: The History of Non-Western Mathematics*. Dordrecht: Springer. URL:
<https://ia800208.us.archive.org/4/items/springer_10.1007-978-94-011-4301-1/10.1007-978-94-011-4301-1.pdf>;

Report from the Workshop on Future Directions in Mathematics, 12-14 October 2011, Institute for Pure and Applied Mathematics, University of California, Los Angeles; URL:
<http://www.ipam.ucla.edu/wp-content/uploads/2014/12/Future-directions-final-report-with-cover.pdf>;

Shives, Matthew S. (2012), Paradigms and Mathematics: A Creative Perspective:
<https://www.maa.org/sites/default/files/images/images/upload_library/46/HOMSIGMAA/Shives_HOM_winner_2013.pdf>;

Rosa, Milton, Ubiratan D'Ambrosio, Daniel Clark Orey, Lawrence Shirley, Wilfredo V. Alangui, Pedro Palhares, Maria Elena Gavarrete (2016), *Current and Future Perspectives of Ethnomathematics as a Program.* Cham:Springer Open. URL:<https://link.springer.com/content/pdf/10.1007%2F978-3-319-30120-4.pdf>;

Blaszczyk, Piotr, Vladimir Kanovei, Karin U. Katz, Mikhail G. Katz, Semen S. Kutateladze, and David Sherry (2016), Toward a history of mathematics focused on procedures: <https://arxiv.org/pdf/1609.04531.pdf>;

Historia Mathematica:
<https://www.mathunion.org/ichm/historia-mathematica>;

Electronic Journal for History of Probability and Statistics:
<http://www.jehps.net/>

PRESENT AND FUTURE

As a part of the unevenly evolving human culture in the world, mathematics is evolving unevenly in different human societies, located over multiple geographical spaces, and passing through multiple social and historical times. Mathematics in the singular for all societies and for all times is a stubborn ideological construct that refuses to notice the historical evolution of the various branches of this discipline. Anthropologists and ethnologists of mathematics continue to document the mathematical history and current practices of the social groups engaged in various levels of hunting-gathering, agricultural and artisanal activities. Mathematics instruction in the state and market sponsored academic institutions of some of the more complex agrarian and industrial societies has attracted some critical attention (Luitel, 2009; Ramanujam and Subramaniam, 2012; Karp and Schubring, 2014). The plurality of interacting academic mathematical cultures (Benthem *et al*, 2006; Friend, 2014; Chakraborty and Friend, 2017) of primary, secondary and tertiary mathematics instruction in the societies currently going through various levels of the first, second, third and fourth... industrial revolutions deserve further systematic critical investigation.

Take the case of instruction of elementary level differential and integral calculus as an example. In most schools, colleges and universities the textbooks and teachers oriented on classical analysis of the nineteenth and twentieth centuries still dominate the scene. In spite of the best efforts of some enthusiasts in some countries [Keisler, 1976; Borovik and Katz, 2012; Kanovei *et al,* 2015] the presence of the nonstandard methods of analyses in the classrooms remains peripheral in many societies, even at the tertiary level. It appears that in future the development of technology will change several aspects of the existing situation, through a process of removal, retraining and renewal of the current generation of the mathematically skilled labour forces of the various industrialized societies, and through the emergence of ever newer man-machine interfaces in the domains of mathematics use, teaching and research.

In the words of Bruno Buchberger: <http://www.risc.jku.at/people/buchberger/>,

Professor of Computer Mathematics at the Research Institute for Symbolic Computation, Johannes Kepler University Linz: <http://www.risc.jku.at/>and, a

Member of the Global Digital Mathematical Library Working Group: <https://blog.wias-berlin.de/imu-icm-panel-wdml/tag/gdml/>:

"[M]athematics is essentially software. In fact from the beginning of mathematics, it was the goal of mathematics to automate problem solving. By systematic and deep thinking, for problems whose solution was difficult in each individual instance, systematic procedures were found that allow to solve each instance without further thinking. In each round of automation in mathematics, the deep thinking on spectra of problem instances is reflected by deep theorems with deep proofs."

"In 20th century, the systematic procedures for spectra of problems became physically tangible as algorithms/software for the universal computer (which itself essentially is a mathematical invention). In 21st century, the rounds of automation in mathematics reach higher and higher levels and move more and more to the meta-level of mathematics, i.e. to the automation of mathematical thinking itself" (Buchberger, 2014b: 9). His considered observations on the future directions of mathematics and of its instruction are available in: Buchberger, 1990, 2003 and 2014a. On the importance of his work for contemporary and future mathematics, computer science and logic, see: Kapur (Ed), 2006; Paule, 2013.

Softwares are written in formal languages. The known history of formal languages may be traced back to the *Aṣṭādhyāyī* [*A Text Containing Eight Chapters*] of Pāṇini [possibly c. 4th century BCE]. This text is an analysis of what its author called *bhāṣā* [language(s) known to him] together with a generative calculus resembling what later on came to be known as an algorithm. The *Aṣṭādhyāyī* played an important role in the rise and development of comparative philology and linguistics since the nineteenth century. It has become an object of investigation in computational linguistics in the twenty-first century (Joshi, 2009).

Karl Marx studied some aspects of the formal languages of elementary algebra, analytical geometry, trigonometry and differential calculus from the texts available to him. Today the nature and history of all formal languages from the time of Pāṇini to the present are accessible to all willing investigators enjoying uninterrupted power supply and net connectivity. Finally, the name and dominant images of Karl Marx first and foremost evoke various associations with a fundamental yearning for equality, justice and human self-emancipation and, of their grotesque miscarriage under the rule of the communist parties of the world. The history of struggles for justice and self-emancipation and, of their serial setbacks is the history of civil society. This history incorporates within it the history of mathematics among other things. Contemporary research has pushed back the history of the interfaces of state, justice and mathematics in complex societies at least to the 21st century BCE (Høyrup, 2009). It is the task of the present and future students of Marx to continue to investigate all mathematics, from the 21 century BCE to the 21st century CE and beyond at both ends, critically and in all their interconnections, in the interests of human self-emancipation. These are my personal observations and views on Karl Marx and mathematics as of now; other perspectives are possible and necessary.

REFERENCES

Alcouffe, Alain (1985), "Marx, Hegel, et le <<calcul>>" ["Marx, Hegel, and the <<calculus>>"], in: *Les Manuscrits Mathématiques de Marx* [*The mathematical manuscripts of Marx*], Paris: Economica. Details at: <http://alain.alcouffe.free.fr/index.php?option=com_content&task=view&id=42&Itemid=2>

____ (1987), "Le calcul différentiel, les mathématiciens et les économistes au XIX°- K. Marx et H. Laurent, lecteurs de J.L. Boucharlat" ["The differential calculus, mathematicians and economists in the nineteenth century: K. Marx and H. Laurent, readers of J.L. Boucharlat"], *Sciences et techniques en perspectives*, 1987, 13: 47-69; URL: <https://hal.archives-ouvertes.fr/hal-01312597/document>; English translation: in part two of the present collection.

Alcouffe, Alain and Julian Wells (2009), Marx, maths, and *MEGA* 2; Paper presented at the 13th Annual Conference of the European Society for the History of Economic Thought, University of Macedonia, Thessaloniki, Greece. URL: <https://mpra.ub.uni-muenchen.de/80535/1/MPRA_paper_80535.pdf>

Alten, Heinz-Wilhelm, Alireza Djafari Naini, Bettina Eick, Menso Folkerts, Hartmut Schlosser, Karl-Heinz Schlote, Heiko Wesemüller-Kock, Hans Wußing (2003), *4000*

Jahre Algebra: Geschichte-Kulturen-Menschen [*4000 Years of Algebra: History-Cultures-People*]; Zweite, aktualisierte und ergänzte Auflage [Second, updated and supplemented edition], 2014, Berlin/Heidelberg: Springer-Verlag.

An Account of the Life and Character of Tofuzzel Hussein Khan, *the* Vakeel, *or,* Ambassador, *of the* Nabob Vizier Assof-ud-Dowlah, *at Calcutta, during the Government of* Marquis Cornwallis, in: *The Asiatic Annual Register for 1803* (1804), Characters: 7, London: Cadell and Davies etc.

Anellis, Irving H. (1996), "Sof'ya Aleksandrovna Yanovskaya's Contributions to Logic and History of Logic", *Modern Logic,* Volume 6, No. 1: 7-36; URL: <https://projecteuclid.org/download/pdf_1/euclid.rml/1204835629>

Antonova, Irina Konstantinovna (1999), "<<Matematicheskie rukopisi>> Marksa (opyt kriticheskovo analiza)" ["*The Mathematical Manuscripts of Marx* (attempt at a critical analysis)"], *Vestnik MGU: Filosofiya* [*Newsletter of the Moscow State University: Philosophy*], 1999, No. 1: 46-62.

____ (2004), *Marksizm vne politiki* [*Marxism beyond politics*], Glava [Chapter] 3: Rekonstruktsia protsessa poznaniya na osnove matematicheskikh rukopisiei Marksa [Reconstruction of the process of cognition on the basis of the mathematical manuscripts of Marx]: 95-136. Moskva: Editorial URSS.

Bag, Amulya Kumar (1966), "Binomial Theorem in Ancient India", *Indian Journal of History of Science,* Volume 1, Issue 1: 68-74.

____ (1979), *Mathematics in Ancient and Medieval India,* Varanasi: Chaukhambha.

Bazhanov, V.A. (2001), "Restoration: S.A. Yanovskaya's path in logic", *History and Philosophy of Logic,* Volume 22, No. 3: 129-133; URL: <http://citeseerx.ist.psu.edu/viewdoc/download?doi=10.1.1.694.2361&rep=rep1&type=pdf>

Benthem, Johan van, Gerhard Heinzmann, Manuel Rebuschi and Henk Visser [Eds.] (2006), *The Age of Alternative Logics: Assessing Philosophy of Logic and Mathematics Today*, Dordrecht: Springer.

Biryukov, Boris Vladimirovich and Biryukova L. G. (2004), "Ludwig Wittgenstein i Sofya Aleksandrovna Yanovskaya: <<Kembridgskii genii>> znakomitsya c sovietskimi matematikami 30-kh godov" ["Ludwig Wittgenstein and Sofya Aleksandrovna Yanovskaya: the "Genius from Cambridge" meets Soviet mathematicians of the 1930s"], *Logicheskie issledovaniya* [*Logical Research*], Volume 11: 46-95. Moskva: Nauka. URL: <https://iphras.ru/uplfile//logic/log11/Li_11_Biryukov_Biryukova.pdf>

Borovik, Alexandre and Mikhail G. Katz (2012), "Who Gave you the Cauchy-Weierstrass Tale? The Dual History of Rigorous Calculus", *Foundations of Science,* Volume 17, Issue 3: 245-276.

Bose, P.N. [Pramatha Nath] (1885), *Centenary Review of the Asiatic Society of Bengal from 1784 to 1883,* Part III. *Natural Science, & c.* Chapter I. §8. Mathematical Sciences of the Hindus, Calcutta: Thacker, Spink and Co.

Buchberger, Bruno (1990), "Should Students Learn Integration Rules?", *SIGSM Bulletin,* Volume 24, Number 1: 10-17; URL: <https://dl.acm.org/citation.cfm?id=1095228&dl=ACM&coll=DL>

____ (2003), White Box / Black Box Principle etc. Invited talk at the Symposium on Mathematics and New Technologies: What to Learn, How to Teach? 10 December,

2003, Universidad Complutense de Madrid, Fundación Ramón Areces, Madrid; URL: <www.risc.jku.at/publications/download/risc_2191/2003-12-10-A.ppt>

____ (2014a), The Future of Mathematics: A Personal View and Comments on Math Education. Keynote address at TIME 2014, Krems, Austria (02 July 2014); URL: <https://journal.ph-noe.ac.at/index.php/resource/article/view/76>

____ (2014b), "Soft Math – Math Soft", in: Hoon Hong and Chee Yap [Eds.], *Mathematical Software – ICMS 2014,* LNCS 8592: 9-15, Berlin etc.: Springer.

Burrow, Reuben (1790), "*A Proof that the* Hindoos *had the* Binomial Theorem", *Asiatick Researches,* Volume the Second: 487-497, Calcutta: Manuel Cantopher and, London: P. Elmsly.

____ (1792), "A Demonstration *of One of the* Hindoo Rules *of* Arithmetic", *Asiatick Researches,* Volume the Third: 145-147, Calcutta: Manuel Cantopher and, London: P. Elmsly.

Chakraborty, Mihir Kumar and Michèle Indira Friend [Eds.] (2017), *Journal of Indian Council of Philosophical Research,* Volume 34, Issue 2, Special Issue: Pluralism in Mathematics, New Delhi: Springer.

Clagett, Marshall (1999), *Ancient Egyptian Science: A Source Book,* Volume Three: *Ancient Egyptian Mathematics.* Philadelphia: American Philosophical Society.

Colebrooke, Henry Thomas [Translator] (1817), *Algebra, with Arithmetic and Mensuration from the Sanscrit of Brahmegupta and Bhascara,* London: John Murray.

Dājī, Bhāu (1865), "Brief Notes on the Age and Authenticity of the Works of Āryabhaṭa, Varāhamihira, Brahmagupta, Bhaṭṭotpala, and Bhāskarāchārya", *Journal of the Royal Asiatic Society of Great Britain and Ireland,* New Series, Volume 1, No. 1/2: 392-418.

Datta, Bibhutibhusan and Avadhesh Narayan Singh (1935, 1938; 1962), *History of Hindu Mathematics: A Source Book,* Parts I and II, Bombay etc.: Asia Publishing House.

Datta, Bibhutibhusan and Avadhesh Narayan Singh, Revised by Kripa Shankar Shukla (1984), "Use of Calculus in Hindu Mathematics", *Indian Journal of History of Science,* 19 (2): 95-104.

Dauben, Joseph Warren (1998), "Marx, Mao and Mathematics: The Politics of Infinitesimals", *Documenta Mathematica*, Extra Volume ICM III, 1998: 799–809. URL: <http://www.emis.ams.org/journals/DMJDMV/xvol-icm/19/Dauben.MAN.html>

____ (2004), "Mathematics and Ideology: The Politics of Infinitesimals", *LlUll,* Volume 27, No. 59: 291-355; URL: <https://dialnet.unirioja.es/servlet/articulo?codigo=2040428>

Demidov, Sergei Sergeevich and Boris Venediktovich Lëvshin (2016), *The case of academician Nikolai Nikolaevich Luzin*, Providence, Rhode Island: American Mathematical Society.

Divakaran, P.P. (2012), "Birth of calculus with special reference to Yuktibhāṣā", *Indian Journal of History of Science*, 47.4 (2012):771–832; URL: <http://www.insa.nic.in/writereaddata/UpLoadedFiles/IJHS/Vol47_4_11_PPDivakaran.pdf>

Dold-Samplonius, Yvonne, Joseph W. Dauben, Menso Folkerts and Benno van Dalen [Eds.] (2002), *From China to Paris: 2000 years transmission of mathematical ideas,* Stuttgart: Franz Steiner Verlag.

Frantz, Constantin (1842), *Die Philosophie der Mathematik. Zugleich ein Beitrag zur Logik und Naturphilosophie* [*The Philosophy of Mathematics. At the Same Time a Contribution to Logic and Natural Philosophy*], Leipzig: Hermann Hartung.

Friend, Michèle Indira (2014), *Pluralism in Mathematics: A New Position in Philosophy of Mathematics,* Dordrecht etc.: Springer.

Gödel, Kurt (1990), Remark on non-standard analysis [1974], *Collected Works,* Volume II, Publications 1938-1974: 311, New York: OUP.

Graham, Loren and Jean-Michel Kantor (2009), *Naming Infinity: A True Story of Religious Mysticism and Mathematical Creativity,* Cambridge, Massachusetts/London, England: The Belknap Press of Harvard University Press.

Gumbel, Emil Julius (1927), "O matematicheskikh rukopisiyakh K. Marksa (Soobschenie)" ["On the mathematical manuscripts of K. Marx (A Communication)"], *Letopisi marksizma* [*Chronicles of Marxism*], 3/1927: 56-60. English translation in: Marx, 2018: ix-xiv.

Gupta, Radha Charan (2002), Chapter 18: India, in: Joseph W. Dauben and Christoph J. Scriba [Eds.], *Writing the History of Mathematics: Its Historical Development:* 307-315, Basel etc.: Birkhäuser Verlag.

Hollings, Christopher (2013), "The Struggle Against Idealism: Soviet Ideology and Mathematics", *Notices of the American Mathematical Society,* Volume 60, Number 11: 1448-1458; URL: <http://www.ams.org/notices/201311/rnoti-p1448.pdf>

Høyrup, Jens (2009), "State, "Justice", Scribal Culture and Mathematics in Ancient Mesopotamia": Sarton Chair Lecture, *Sartoniana,* 22: 13-45; URL: <https://rucforsk.ruc.dk/ws/files/4344168/Hoyrup_2009_j__State___Justice___Scribal_Culture_and_Mathematics_in_Ancient_Mesopomia.pdf>

Imhausen, Annette (2016), *Mathematics of Ancient Egypt: A Contextual History,* Princeton/Oxford: Princeton University Press.

Joshi, Shivram Dattatray (2009), Background of the Aṣṭādhyāyī; Keynote Speech at the Third International Sanskrit Computational Linguistics Symposium, 15-17 January 2009, Department of Sanskrit Studies, University of Hyderabad, Hyderabad, India. Available in pages 1-5 of the Proceedings of the Symposium: <http://www.springer.com/la/book/9783540938842>

Kanovei, Vladimir, Karin U. Katz, Mikhail G. Katz and Mary Schaps (2015), "Proofs and Retributions, Or: Why Sarah Can't *Take* Limits", *Foundations of Science,* Volume 20, Issue 1: 1-25.

Kapur, Deepak [Ed.] (2006), *Logic, Mathematics and Computer Science: Interactions in Honor of Bruno Buchberger (60th Birthday), Journal of Symbolic Computation,* Volume 41, Issues 3-4, pages 255-512 (March-April 2006); URL: <https://www.sciencedirect.com/journal/journal-of-symbolic-computation/vol/41/issue/3-4>

Karp, Alexander and Gert Schubring [Eds.] (2014), *Handbook on the History of Mathematics Education,* New York: Springer.

Katz, Victor J. *et al* [Eds.] (2007), *The Mathematics of Egypt, Mesopotamia, China, India, and Islam: A Sourcebook,* Princeton/Oxford: Princeton University Press.

____ (2016), *Sourcebook in the Mathematics of Medieval Europe and North Africa,* Princeton/Oxford: Princeton University Press.

Keisler, Howard Jerome (1976), *Elementary Calculus: An Approach Using Infinitesimals.* Boston: Prindle, Weber and Schmidt. *Elementary Calculus: An Infinitesimal Approach* (2000). Revised in August 2017: <https://www.math.wisc.edu/~keisler/calc.html>

Keller, Agathe (2006), *Expounding the Mathematical Seed: A Translation of Bhāskara I on the Mathematical Chapter of the Āryabhatīya,* Volume 1: The Translation; Volume 2: The Supplements. Basel etc.: Birkhäuser Verlag.

Kennedy, Hubert (1977), Marx, Peano and Differentials [a paper read at the 15th International Congress of the History of Science, Edinburg, August 1977]. URL: <http://www.autodidactproject.org/other/sn-peano1.html>

____ (2006), *Negation of Negation: Karl Marx and the Differential Calculus.* A collection of articles originally published in the years 1976-1982, Concord, California: Peremptory Publications.

Khan, Iqbal Ghani (2003), "The Awadh scientific renaissance and the role of the French: c. 1750-1820", *Indian Journal of History of Science,* 38.3 (2003): 273-301; URL: <http://124.108.19.235:12000/jspui/bitstream/123456789/10357/1/Vol38_3_5_IGKhan.pdf>

Körle, Hans-Heinrich (2015), *Infinite Series in a History of Analysis: Stages up to the Verge of Summability,* Berlin/Boston: Walter De Gruyter GmbH.

Küng, Guido (1962), "Bibliography of Soviet work in the field of mathematical logic and the foundations of mathematics, from 1917 – 1957", *Notre Dame Journal of Formal Logic,* Volume III, Number 1 (1962): 1-40; URL: <https://projecteuclid.org/download/pdf_1/euclid.ndjfl/1093957057>

Kurbjuweit, Dirk (1993), "Karl Marx ist tot. Friedrich Engels ist tot. Die Mega lebt. Die MEGA-STARS"["Karl Marx is dead. Friedrich Engels is dead. The MEGA lives. The MEGA-STARS"], *ZEIT ONLINE,* 7. Mai 1993, 8:00 Uhr; Aktualisiert am 18. Oktober 2013, 14:04 Uhr; available at: <https://www.zeit.de/1993/19/die-mega-stars/komplettansicht>

Kusraev, Anatolii Georgievich and Semën Samsonovich Kutateladze (1994), *Nonstandard Methods of Analysis,* Part I: Infinitesimal Analysis, Chapter I: Excursus into the history of mathematical analysis, 1.2 Karl Marx on Mysterious Differential Calculus. Dordrecht: Springer. It is an English translation of: *Nestandartnye metody analiza* (1990), Novosibirsk: Nauka, by L. Iouzina-Tarvainen.

Leckey, Colum (1995), "David Riazanov and Russian Marxism", *Russian History,* Volume 22, No. 2: 127-153; URL: <https://www.academia.edu/21306389/David_Riazanov_and_Russian_Marxism>

Levin, Vitaly (2017), "Sofya Aleksandrovna Yanovskaya – K 120-letiyu so dnya rozhdeniya" ["Sofya Aleksandrovna Yanovskaya – On the Occasion of her 120-th Birth Anniversary"], *Logiko-Filosofskie Studii* [*Logico-philosophical Studies*], T. 15, NO. 1: 44-55; URL: <http://ojs.philosophy.spbu.ru/index.php/lphs/article/view/502/494>

Lorentz, George Gunter [Rudolfovich] (2002), "Mathematics and Politics in the Soviet Union from 1928 to 1953", *Journal of Approximation Theory,* 116: 169-223; URL: <https://www.emis.de/classics/HAT/fpapers/lorentzussr.pdf>

Luitel, Bal Chandra (2009), *Culture, worldview and transformative philosophy of mathematics education in Nepal: a cultural-philosophical inquiry.* Doctoral Dissertation, presented at the Curtin University of Technology, Perth, Australia; URL: <https://espace.curtin.edu.au/handle/20.500.11937/682>

Marks, Karl (1968), *Matematicheskie rukopisi* [*Mathematical Manuscripts*], Edited by Sofya Aleksandrovna Yanovskaya, Moskva: Nauka.

Martzloff, Jean-Claude (1987), *Histoire des mathématiques chinoises*. Paris: Masson. Revised and augmented English translation (1997), *A History of Chinese Mathematics*; Translator: Dr. Stephen S. Wilson, Berlin/Heidelberg: Springer.

Marx, Karl (1994a), *Mathematical Manuscripts* [A complete English translation of Marks, 1968] together with a Special Supplement: *Marx and Mathematics*; translator of the manuscripts and editor of the Special Supplement: Pradip Baksi, Kolkata: Viswakos.

____ (1994b) *Gonit Bishayak Pandulipisamuha, o bishesh songjojani: Marx o Gonit*; Bengali translation of: K. Marks, 1968, together with a Special Supplement: *Marx and Mathematics*, by Pradip Baksi, Kolkata: Viswakos.

____ (2018) Revised second edition of Marx, 1994a, [without the Special Supplement, which has grown into the present volume], Delhi: Aakar Books.

Marx, Karl und Friedrich Engels (2003), *Gesamtausgabe,* II/14: *Manuskripte und redaktionelle Texte zum dritten Buch des "Kapitals". 1871 bis 1895* [*Manuscripts and editorial texts for the third book of the "Capital". 1871-1875*]. Berlin: De Gruyter Akademie Forschung. Inhalt und Einführung [Contents and Introduction]: <http://mega.bbaw.de/struktur/abteilung_ii/ii-14>; <http://mega.bbaw.de/struktur/abteilung_ii/dateien/mega_II-14_inhalt-einf.pdf>

Marxhausen, Thomas (2006), ">>MEGA – MEGA<< und kein Ende" [">>MEGA – MEGA<< and no end"], *UTOPIE kreativ*, H. 189/190 (Juli/August 2006), S. 596-617; URL: <https://www.rosalux.de/fileadmin/rls_uploads/pdfs/189Marxhausen.pdf>

Medvedev, Feodor Andreevich (1987), "Nestandartny analiz i istoriya klassicheskovo analiza" ["Nonstandard analysis and the history of classical analysis"], in Panov (1987): 75-84; English translation: in part three of the present collection.

Molodshii, Vladimir Nikolaevich (1983), "<<Matematicheskie rukopisi>> K. Marksa i razvitie istorii matematiki v SSSR" ["Marx's "Mathematical Manuscripts" and development of history of mathematics in the USSR"], *Voprosy istorii estestvoznaniya i tekhniki,* [*Questions of History of Natural Science and Technology*] 1983, No. 2: 29-34; English translation: in part two of the present collection.

Mosolov, Vladimir Gavrilovich (2010), *IMEL–tsitadel' partiinoi ortodoxii: iz istorii Instituta marksizma-leninizma pri TSK KPSS, 1921-1956* [*IMEL – Citadel of party orthodoxy: from the history of the Institute of Marxism-Leninism of CC CPSU, 1921-1956*], Moskva: Novy khronograf.

Nepeyvoda, Nikolai Nikolaevich (1987), "Stanovlenie poniatia konstruktivnosti v matematike" ["Emergence and development of the concept of constructivisability in mathematics"], in Panov (1987): 219-229; English translation: in part three of the present collection.

Panov, Mikhail Ivanovich [Contributing Editor] (1987), Sbornik statei: *Zakonomernosti razvitiya sovremennoi matematiki: Metodologicheskie aspekty* [Collection of papers:

Regularities in the development of modern mathematics: Methodological aspects], Moskva: Nauka.

Paule, Peter [Ed.] (2013), *Mathematics, Computer Science and Logic – A Never Ending Story: The Bruno Buchberger Festschrift,* Cham etc.: Springer.

Pingree, David (1970-1994), *Census of the Exact Sciences in Sanskrit,* Series A, Volumes 1-5. Philadelphia: American Philosophical Society. URL: <https://archive.org/details/PingreeCESS>

Ponzio, Augusto (2014), "The Semiotics of Karl Marx: A Historical and Theoretical Excursus through the Sciences of Signs in Europe", *Chinese Semiotic Studies,* Volume 10, Issue 2 (2014): 195-214.

Przhesmitsky, Vasily Ivanovich (1985), "Operativnyi logicheskii apparat rabotayuschii v *Kapitale* i *Matematicheskikh rukopisiyakh* K. Marksa" ["On the operational logical apparatus working in Karl Marx's *Capital* and *Mathematical Manuscripts"*], *Voprosy dialekticheskoi logiki: Printsipy i formy myshleniya* (materialy postoyanno deistvuyuschevo simpoziuma po dialekticheskoi logike) [*Problems of dialectical logic: Principles and forms of thought* (materials of the permanently acting symposium on dialectical logic)], 1985: 70-81, Moskva: A N SSSR, Institut Filosofii.

Ramanujam, R. and K. Subramaniam (2012), *Mathematics Education in India: Status and Outlook,* Mumbai: Homi Bhaba Centre for Science Education; URL: <http://mathedu.hbcse.tifr.res.in/wp-content/uploads/2014/01/INP-Book_Mathematics-Edu-in-India_2012_KS-RR.pdf>

Ramchundra [Ramchandra Lal] (1850), *A Treatise on Problems of Maxima and Minima solved by Algebra,* Calcutta: P.S. D'Rozario and Co. Reprinted with Editor's Preface by Augustus De Morgan (1859), London: WM.H. Allen & Co.

Rashed, Roshdi (1994), *The development of Arabic mathematics: between arithmetic and algebra,* Dordrecht: Springer. English translation by A.F.W. Armstrong of: Roshdi Rashed (1984), *Entre Arithmetique et Algebre. Recherches sur l'Histoire des Mathématiques Arabes,* Paris: Les Belles Lettres.

____ (2015), *Classical Mathematics from al-Khwārizmī to Descartes,* London and New York: Routledge. English translation by Michael H. Shank of his *D'al-Khwārizmī à Descartes*: *Études sur l›histoire des mathématiques classiques,* Paris: Hermann, 2011.

Rokityansky, Yakov Grigorievich (2008), "Iz biografii akademika D.B. Riazanova: razgrom Instituta K. Marksa i F. Engelsa (Mart 1931g.)" ["From the Biography of Academician D.B. Riazanov: the Rout of the Institute of K. Marx and F. Engels (March 1931)"], *Otechestvennye arkhivy* [*National Archives*], No. 4: 10-23; URL: <http://portal.rusarchives.ru/publication/ryazanov.shtml>

Rosenfeld, Boris (1996), "Reminiscences of S.A. Yanovskaya", *Modern Logic,* Volume 6, No. 1: 73-76.

Roy, Ranjan (2011), *Sources in the Development of Mathematics: Infinite Series and Products from the Fifteenth to the Twenty-first Century,* Cambridge etc.: Cambridge University Press.

Ruzavin, Georgi Ivanovich (1987), "Novy strukturny podkhod k matematike i nekotorye problemy eo metodologii" ["The new structural approach in mathematics and some of its methodological problems"], in Panov, 1987: 155-169; English translation: in part three of the present collection.

Sarma, K.V. [Ed.] (2008), *Gaṇita-Yukti-Bhāṣā [Rationales of mathematical astronomy] of Jyeṣṭhdeva (c. 1500-1575)*, in Two Volumes: Volume I: Mathematics; Volume II: Astronomy, New Delhi: Hindustan Book Agency and London: Springer-Verlag.

Scriba, Christoph J. and Peter Schreiber (2001), *5000 Jahre Geometrie: Geschichte, Kulturen, Menschen,* Berlin/Heidelberg: Springer-Verlag. An English translation of its updated and supplemented 3rd edition of 2010 by Jana Schreiber (2015), *5000 Years of Geometry: Mathematics in History and Culture,* Basle: Springer.

Seneta, Eugene (2004), "Mathematics, religion, and Marxism in the Soviet Union in the 1930s", *Historia Mathematica,* 31 (2004): 337-367. URL: <https://www.sciencedirect.com/science/article/pii/S0315086003000466>

Sengupta, Prabodh Chandra (1932), "Infinitesimal calculus in Indian mathematics – its origin and development", *Journal of the Department of Letters* (University of Calcutta), Volume XXII, 5th article: 1-17, Calcutta: Calcutta University Press.

Serman, Pascal (2013), "Karl Marx et le calcul infinitésimal" ["Karl Marx and infinitesimal calculus"], *Repères-IREM.* Num. 91. pp. 93-103; URL: <http://numerisation.irem.univ-mrs.fr/WR/IWR13008/IWR13008.pdf>

Shastri, Bapu Deva (1858), "Bhāskara's knowledge of the Differential Calculus", *Journal of the Asiatic Society of Bengal,* Volume XXVII, No. III: 213-216.

Sinkevich, Galina (2014), "Istoriografiya matematicheskovo analiza" ["Historiography of mathematical analysis"], v sbornike [in the collection]: *Matematicheskoe modelirovanie, chislennye metody i komplekcy programm [Mathematical modeling, numerical methods and program complexes]*: 3-22. Vypusk [Issue] 20, St. Petersburg: Izdatel'stvo SPbGASU.

Smorynski, Craig (2017), *MVT: A Most Valuable Theorem,* Chapter 3: The Mean Value Theorem, 3.2: Precursors to the Mean Value Theorem, 3.2.3: The Analytic Approach in Mediæval India: 188-215. Cham: Springer International.

Sonar, Thomas (2011), *3000 Jahre Analysis [3000 Years of Analysis]*, Berlin/Heidelberg: Springer.

Sperl, Richard (2005), "Marx-Engels-Editionen" ["Marx-Engels-Editions"], S. 329-360, in: *Editionen zu deutschsprachigen Autoren als Spiegel der Editionsgeschichte[Editions of German Speaking Authors as a Mirror of History of Editions]*, Hrsg. Rüdiger Nutt-Kofoth und Bodo Plachta, Tübingen: Max Niemeyer.

____ (2016), "Jede Edition ist ein kind ihrer Zeit": Zur Editionsgeschichte des literarischen Nachlasses von Karl Marx und Friedrich Engels ["Every Edition is a Child of its Time": On the History of Editions of the Literary Legacy of Karl Marx and Friedrich Engels]; URL: <https://marx200.org/en/forewords/jede-edition-ist-ein-kind-ihrer-zeit>

Spottiswoode, William (1860), "Note on the supposed discovery of the Principle of the Differential Calculus by an Indian Astronomer", *The Journal of the Royal Asiatic Society of Great Britain and Ireland,* Volume the Seventeenth, Article VII: 221-222.

Tytler, John (1820), "Essay on the Binomial Theorem; as known to the Arabs", *Asiatick Researches,* Volume the Thirteenth: 456-466, Calcutta: The Calcutta Gazette Office.

____ (1832), "An Essay on the extraction of the Roots of Integers, as practised by the Arabs", *Asiatic Researches,* Volume XVII: 51-168, Calcutta: The Bengal Military Orphan Press.

____ (1837), "Analysis and Specimens of a Persian Work on Mathematics and Astronomy", *The Journal of the Royal Asiatic Society of Great Britain and Ireland,* Volume the Fourth: 254-272, London: John W. Parker.

Uspensky, Vladimir Andreeivich (1987), "Sem' razmyshlenii na temi filosofii matematiki" ["Reflections on seven themes of philosophy of mathematics"], in Panov (1987): 106-155; English translation: in part three of the present collection.

Vasiliev, V. V. (2017), "Moskovskie adresa Wittgensteina" ["The Moscow addresses of Wittgenstein"], *Voprosy filosofii* [*Problems of Philosophy*], 2017, No. 6: 39-51; URL: http://vphil.ru/index.php?option=com_content&task=view&id=1664&Itemid=52

____ (2017b), Moskva i Wittgenstein: den' za dneom [Moscow and Wittgenstein: a Day by Day Chronicle]. URL: <http://hardproblem.ru/vitgenshtejn-v-moskve-den-za-dnem/>; <http://hardproblem.ru/moskva-i-vitgenshtejn-den-za-dnem-polnoerassledovanie/>

Vogt, Annette (1995), "Emil Julius Gumbel (1891-1966): der erste Herausgeber der mathematischen Manuskripte von Karl Marx" ["Emil Julius Gumbel (1891-1966): The first editor of the mathematical manuscripts of Karl Marx"], *MEGA-Studien* (Amsterdam), 1995/2: 26-36; English translation: in part one of the present collection.

Vucinich, Alexander (1999), "Mathematics and Dialectics in the Soviet Union: The Pre-Stalin Period", *Historia Mathematica,* 26 (1999): 107-124. URL: <https://www.sciencedirect.com/science/article/pii/S0315086099922292>

____ (2000), "Soviet Mathematics and Dialectics in the Stalin Era", *Historia Mathematica,* 27(2000): 54-76.URL: <https://www.sciencedirect.com/science/article/pii/S0315086099922619>

____ (2002), "Soviet Mathematics and Dialectics in the Post-Stalin Era: New Horizons", *Historia Mathematica,* 29 (2002): 13-39.URL: <https://www.sciencedirect.com/science/article/pii/S0315086001923298>

Weil, André (1984), *Number Theory: An approach through history from Hammurapi to Legendre,* Boston etc.: Birkhäuser.

Whish, Charles Matthew (Read in 1832 and published in 1834), "On the Hindū Quadrature of the Circle, and the infinite series of the proportion of the circumference to the diameter exhibited in the four Śāstras, the Tantra Sangraham, Yucti Bhāshā, Carana Paddhati, and Sadratnamāla," *Transaction of the Royal Asiatic Society of Great Britain and Ireland,* Vol. 3, No. 3 (1834): 509-523; URL: <https://www.jstor.org/stable/pdf/25581775.pdf>

Wilkinson, Lancelot (1837), "Discovery of the Rekha Ganita, a translation of the Elements of Euclid into Sanskrit…", *The Journal of the Asiatic Society of Bengal,* Volume VI, Part II, July to December 1837: 938-948 + Diagrams, Calcutta: The Society's Office.

Wußing, Hans (2008), *6000 Jahre Mathematik: Eine kulturgeschichtliche Zeitreise – 1. Von den Anfängen bis Leibniz und Newton* [*6000 Years of Mathematics: A Cultural-Historical Journey through Time – 1. From the Beginnings to Leibniz and Newton*], Berlin/Heidelberg: Springer-Verlag.

____ (2009), *6000 Jahre Mathematik: Eine kulturgeschichtliche Zeitreise –2. Von Euler bis zur Gegenwart* [*6000 Years of Mathematics: A Cultural-Historical Journey through Time – 2. From Euler to the Present*], Berlin/Heidelberg: Springer-Verlag.

Xu, Yibao (2013), Dialectics of Numbers: Marxism, Maoism, and the Calculus of Infinitesimals, Chapter 5: 165-196, in: *Mr. Science and Chairman Mao's Cultural Revolution: Science and Technology in Modern China,* Edited by Chunjuan Nancy Wei and Darryl E. Brock, Lanham etc.: Lexington Books.

Yanovskaya, Sofya Aleksandrovna (1930), "Ocherednye zadachi matematikov-marksistov" ["Immediate tasks of the mathematician-marxists"], *Pod znamenen marksizma* [*Under the Banner of Marxism*], 1930, No. 5: 88-94.

____ (1931), "Idealizm v sovremennoi filosofii matematiki" ["Idealism in contemporary philosophy of mathematics"], *Na bor'bu za materialisticheskuyu dialektiku po matematike* [*On the struggle for materialist dialectics in mathematics*], Sbornik statei [Collection of papers]. Red. [Ed.] E. Kol'man; Moskva/Leningrad: Gosudrastvennoe nauchno-tekhnicheskoe izdatel'stvo.

____ (1933), "O matematicheskikh rukopisiyakh K. Marksa" ["On the mathematical manuscripts of K. Marx"], *Pod znamenen marksizma* [*Under the Banner of Marxism*], 1933, No. 1: 74-115.

____ Red. [Ed.](1936), *Sbornik statei po filosofii matematiki* [*Collection of papers on philosophy of mathematics*], Moskva: Uchpedgiz.

____ (1968), Predislovye [Preface], K. Marks, *Matematicheskie rukopisi* [*Mathematical Manuscripts*], Moskva: Nauka. Reprinted in: Yanovskaya, S.A. (1972), *Metodologicheskie problemy nauki* [*Methodological problems of science*]: 12-33, Moskva: Mysl.

Yushkevich, Adolf-Andrei Pavlovich (1987), "Matematika i eo istoriya v retrospektive" ["Mathematics and its history in retrospective"], in Panov (1987): 28-74; English translation: in part three of the present collection.

Acknowldgements

I am once more indebted to Sergei Serebriany, Director, E.M. Meletinsky Institute for Advanced Studies in the Humanities, Russian State University for the Humanities at Moscow. He kindly sent me a scanned copy of a paper on Marx's mathematical manuscripts written by Sofya Aleksandrovna Yanovskaya in 1933.

I express my sincere thanks to Annette Vogt, Senior Research Scholar, Max-Planck Institut für Wissenschaftsgeschichte and Honorary Professor, Humboldt-Universität zu Berlin for sending me a copy of her paper on Emil Julius Gumbel, the first editor of Marx's mathematical manuscripts, published in 1995, a few years ago and, for giving me the permission to include my translation of that paper in the present collection.

I remain grateful to Alain Alcouffe, Université Toulouse 1 Capitole, Economics, Emeritus, for kindly reminding me about the work done by Irina Konstantinovna Antonova for *MEGA*^{2} I/28 and, for the information related to the subsequent developments culminating in the postponement *sine die* of the work on Marx's mathematical manuscripts within the current project of *MEGA*^{2}.

I also remain indebted to him for his contribution to this volume. It is his English rendering of his paper titled "The differential calculus, mathematicians and economists in the nineteenth century: K. Marx and H. Laurent, readers of J.L. Boucharlat", originally published in French in 1987.

Finally, I remain grateful to Kaushal Kumar Saxena for his decision to publish this volume.

The usual disclaimers apply everywhere.

Pradip Baksi

Kolkata
19 April 2019

Introduction, 1994

This special supplement to Marx's *Mathematical Manuscripts* has grown out of an attempt to write a new preface to it. The supplement has three parts.

PART ONE: HISTORY, contains materials pertaining to the history of evolution of Marx's mathematical investigations, to that of the work leading to the publication of their results and, a bibliography, listing the majority of existing publications the field. The bibliography is not exhaustive.

PART TWO: INVESTIGATIONS, contains four articles inspired by Marx's mathematical manuscripts. Three of them are translations from Russian. They relate Marx's mathematical investigations respectively to: the history of analysis, the study of history of mathematics in the erstwhile USSR and, the logic of Marx's *Capital*. The fourth article provides an outline of the problem of situating Marx's mathematical manuscripts in the history of ideas as a whole.

Parts one and two reflect the past—the work that has already been done. The question arises: where do we go from here? Marx conducted his investigations in the 19th century, basing himself mainly upon the developments in analysis upto the end of 18th-beginning of 19th centuries (e.g., upto the time of Lagrange). We are living in the last decade of the 20th century. In Marx's lifetime, and even a couple of decades after his death—right upto the first decade of this century—practically, there existed *only one* (now called the classical) trend in mathematical analysis, but now there are *many* (classical, intuitionist, constructivist and non-standard—some of them overlap and, branch out into multiple sub-trends). Indeed, the fact that even to-day we speak of mathematics in the singular, reflects a particular view of it. We must also train ourselves to speak, in the plural, of the mathematics*es*. [This graphic barbarism has been introduced to jolt the reader out of the prevalent singular use of the word *mathematics*, which is morphologically both plural and singular in English. In Bengali and Russian we have greater graphic clarity: *gonit* > *gonitsamuha*; *matematika* > *matematiki*. The grammatically correct English usage would be 'many mathematics' or 'plural mathematics'.] The ontological and epistemological consequences of these more recent developments in the history of mathematics are profound: that which was considered to be one has also revealed itself as *many*. The entire problematique of truth and certainty has entered into an era of radical reconstruction all over again. At long last, the strongest bastion of theoretical dogmatism—the monopoly of the classical mathematical paradigm—is crumbling before our own eyes. It is clear, that is why, that there is no point in beginning our investigations from just where Marx left his work unfinished—when he died in 1883. To-day, even a survey of the post-Lagrange developments must base itself upon the contemporary attainments, the frontiers of which are being extended daily, hourly.

PART THREE: PLURAL MATHEMATICS, has been planned to provide the reader of this volume with a perspectival update on the relevant developments. Owing to reasons beyond the control of the present author here the cut year is 1987—when all the five articles included in this part were published in Russian, as part of the proceedings of a symposium on "The Regularities and Modern Tendencies of the Development of Mathematics", held in September 1985, in Obninsk.

All the articles included in this supplement are, to my knowledge, being published in English for the first time. All translations from Russian are mine. It will be noted that all the articles herein included—save my own—are translations from Russian. A number of factors determined this choice. A vast amount of relevant work exists in Russian, and our readers are generally unaware of them—this situation requires correction. Where a collective effort is needed, I was constrained to work single-handedly, without any kind of institutional support. I have deliberately excluded the materials already available in English, since our leaders have greater access to them, thanks to our historical contacts with the English-using world. Thus, a very relevant article by *John Kadvany*, A Mathematical Bildungsroman // *History and Theory*, 1, 1989, pp. 25-42, which should otherwise have been reprinted in part three of this supplement, remains excluded. <This article has been superseded by: John Kadvany (2001), *Imre Lakatos and the Guises of Reason*, Durham and London: Duke University Press.> I strongly recommend it for any reader of Marx and Mathematics. I have heard that the Italian publications on Marx's mathematical manuscripts are really very good; but I do not read Italian; in fact, apart from Russian I do not read any other European language. This personal limitation has also contributed to the inadequacies of this supplement. Let us hope that in future someone will make the other relevant materials accessible to us.

The limitations of the present volume and of this supplement will be overcome with the publication of a better and complete edition of all the mathematical manuscripts of Marx (some 400 pages of them still remain unpublished), as well as with the publication of newer and newer studies on them, executed with ever greater competence; but more importantly, we must join hands in opening up new frontiers in mathematical theory and practice and, in the cognate disciplines (e.g., in informatics) and technologies, and thus carry forward the critical and transformative spirit embodied in the mathematical manuscripts of Karl Marx .

Pradip Baksi

Calcutta, June 15, 1993.

Part One

History

On the History of Emergence, Development, Reception and Editing of the Mathematical Manuscripts of Marx

EXCERPTS FROM LETTERS

Marx's mathematical investigations have been discussed in a number of letters of Marx and Engels. English translations of the relevant sections of some of the hitherto published relevant letters are being reproduced here.

Marx to Engles

In Manchester

[London,] 11 January [1858]

Dear Frederick,

*　　*　　*

In elaborating the PRINCIPLES of economics[1]* I have been so damnably held up by errors in calculation that in DESPAIR I have applied myself to a rapid revision of algebra. I have never felt at home with arithmetic. But by making a detour via algebra, I shall quickly get back into the way of things.

*　　*　　*

your

K.M.

[*MECW*, 40, 244]

* For the notes see pp. 13-14.

Marx to Engels

In Manchester

London, 6 July 1863

Dear Engels,

* * *

... My spare time is now devoted to differential and integral calculus. Apropos, I have a superfluity of works on the subject and will send you one, should you wish to tackle it. I should consider it to be almost essential to your military studies. Moreover, it is a much easier branch of mathematics (so far as mere technicalities are concerned) than, say, the more advanced aspects of algebra. Save for a knowledge of the more ordinary kind of algebra and trigonometry, no preliminary study is required except a general familiarity with conic sections.

* * *

your

K.M.

[*MECW*, 41, 484]

Engels to Friedrich Albert Lange[2]

In Duisburg

Manchester, 29 March 1865
7 Southgate

Dear Sir,

* * *

There is a remark about old Hegel which I cannot let pass without comment: you deny him any deeper knowledge of the mathematical sciences . Hegel knew so much mathematics that none of his disciples was capable of editing the numerous mathematical manuscripts that he left behind[3]. The only man who, to my knowledge, has enough understanding of mathematics and philosophy to be able to do so is Marx.

* * *

Yours very respectfully
Friedrich Engels

[*MECW*, 42, 138]

Marx to Engels

In London

[Manchester,] 31 May 1873
25 Dover Street

Dear Fred,

* * *

I have been telling Moore[4] about a problem with which I have been racking my brains for some time now. However, he thinks it is insoluble, at least *pro tempore*, because of the many factors involved, factors which for the most part have yet to be discovered. The problem is this: you know about those graphs in which the movements of prices, discount rates, etc., etc., over the year etc., are shown in rising and falling zigzags. I have variously attempted to analyse crises by calculating these UPS AND DOWNS as irregular curves and I believed (and still believe it would be possible if the material were sufficiently studied) that I might be able to determine mathematically the principal laws governing crises[5]. As I said, Moore thinks it cannot be done at present and I have resolved to give it up FOR THE TIME BEING.

* * *

your
K.M.

[*MECW*, 44, 504]

Engels to Marx

In London

August 18, 1881

Dear Moor[6],

... yesterday I found the courage at last to study your mathematical manuscripts[7] even without reference books, and I was pleased to find that I did not need them. I compliment you on your work. The thing is as clear as daylight, so that we can wonder enough at the way the mathematicians insist on mystifying it. But this comes from the one-sided way these gentlemen think . To put $\frac{dy}{dx} = \frac{0}{0}$, firmly and point blank, does not enter their skulls. And yet it is clear that $\frac{dy}{dx}$ can only be the pure expression of a completed process it the last trace of the *quanta* x and y has disappeared, leaving the expression of the preceding process of their change without any quantity.

You need not fear that any mathematician has preceded you here. This kind of differentiation is indeed much simpler than all others, so that just now I applied it myself to derive a formula I had suddenly lost, confirming it afterwards in the usual way. The procedure must have made the greatest sensation, especially, as is clearly proved, since the usual method of neglecting *dxdy* etc. is *positively false*. And that is the special beauty of it; only if $\frac{dy}{dx} = \frac{0}{0}$, *is* the mathematical operation absolutely correct.

So old Hegel guessed quite correctly when he said that differentiation had for its basic condition, that the variables must be raised to different powers, and at least one of them to at least the second, or 1/2, power[8]. Now we also know why.

If we say that in $y = f(x)$ the x and y are variables, then this claim has no further consequences, as long as we do not move on, and x and y are still *pro tempore*, in fact constants. Only when they really change, i.e., *inside the function*, do they indeed become variables, and only then can the relation still hidden in the original equation reveal itself—not the relation of the two magnitudes but of their variability. The first derivative $\frac{\Delta y}{\Delta x}$ shows this relation as it happens in the course of a real change, i.e., in each given change; the completed derivative $\frac{dy}{dx}$ shows it in its generality, pure, and hence we can come from $\frac{dy}{dx}$ to each $\frac{\Delta y}{\Delta x}$, while the latter itself only covers the special case. However, to pass from the special case to the general relationship, the special case must be sublated (*aufgehoben*) as such.

Hence, after the function has passed through the process from x to x' [in the notation now in use: to x_1—Ed.] with all its consequences, x' can be allowed calmly to become x again; it is no longer the old x, which was variable in name only; it has passed through *actual change*, and the result of the change remains, even if we again sublate (*aufheben*) it.

At last we see clearly, what mathematicians have claimed for a long time, without being able to present rational grounds, that the (derivative or) *differential quotient* is the original, the differentials *dx* and *dy* are derived: the derivation of the formulae demands that both so-called irrational [here "irrational" means "non-rational"—Ed.] factors stand at the same time on one side of the equation, and only if you put the equation back into its first form $\frac{dy}{dx} = f'(x)$, as you can, are you free of the irrationals [i.e. "non-rationals"—Ed.] and instead have their rational expression.

The thing has taken such a hold of me that it not only goes round my head all day, but last week in a dream I gave a chap my shirt-buttons to differentiate, and he ran off with them.

Yours

F.E.

[*Mathematical Manuscripts of Karl Marx*, London, 1983, XXVII-XXVI1L]
[For another translation of this letter, see: *MECW*, 46, 130-132]

Engels to Marx

In Ventnor

London, November 21, 1882

Dear Moor,

... enclosed a mathematical essay by [Samuel] Moore. The conclusion that "the algebraic method is only the differential method disguised" refers of course only to his own method of geometrical construction and is pretty correct there too. I have written to him that you place no value on the way the thing is represented in geometrical construction, the application to the equation of curves being quite enough. Further, the fundamental difference between your method and the old one is that you make x change to x' [in the notation now in use: to x_1—Ed.], thus making them really vary, while the other way starts from $x + h$, which is always only the sum of two magnitudes, but never the variation of a magnitude. Your x therefore, even when it has passed through x' and again became the first x, is still other than it was; while x remains fixed the whole time, if h is first added to it and then taken away again. However, every graphical representation of the variation is necessarily the representation of the *completed* process, of the *result*, hence of a quantity which became constant, the line x; its supplement is represented as $x + h$, two pieces of a line. From this it already follows that a graphical representation of how x' again becomes x, is impossible ...

Your

F.E.

[*Mathematical Manuscripts of Karl Marx*, London, 1983, p. XXIX]
[For another translation of this letter, see: *MECW*, 46, 378-379]

Marx to Engels

In London

November 22, 1882
1, St. Boniface Gardens,
Ventnor.

Dear Fred,

... Sam, as you saw immediately, criticises the analytical method applied by me by just pushing it aside, and instead busies himself with the geometrical application, about which I said not one word. In the same way, I could get rid of the development of the proper so-called differential method—beginning with the mystical method of Newton and Leibniz, and then going on to the rationalistic method of d'Alembert and Euler, and finishing with the strictly algebraic method of Lagrange (Which, however, always begins from the same original basic outlook [as that of] Newton-Leibniz)—I could get rid of this whole historical development of analysis by saying that *practically* nothing essential has changed in the geometrical application of the differential calculus, i.e., in the geometrical representation.

The Sun is now shining, so the moment for going for a walk has come, so no more *pro nunc* of mathematics, but I'll come back later to the differential methods occasionally in detail ...

Yours

K.M.

[*Mathematical Manuscripts of Karl Marx*, London, 1983, p. XXX]
[For another translation of this letter, see: *MECW*, 46, 380]

EXCERPTS FROM REMINISCENCES

Marx's friends and comrades have mentioned his mathematical studies in their reminiscences or elsewhere. English translations of the relevant portions of some such writings are being published here.

Engels' Speech At Marx's Funeral (Excerpts)

*** *** ***

... in every single field, wherever Marx has conducted investigations and, it goes without saying that the fields of his investigations were many and variagated, and no where did he tackle the task of investigation superficially—in every field, even in mathematics, he has obtained independent results.

*** *** ***

17 March 1883
[*MECW*, 24: 468]

From The Preface To The Second Edition Of Anti-Dühring

*** *** ***

Marx and I were pretty well the only people to rescue conscious dialectics from German idealist philosophy and apply it in the materialist conception of nature and history. But a knowledge of mathematics and natural science is essential to a conception of nature which is dialectical and at the same time materialist. Marx was well versed in mathematics, but we could keep up with the natural sciences only piecemeal, intermittently and sporadically.

*** *** ***

... to me there could be no question of building the laws of dialectics into nature, but of discovering them in it and evolving them from it.

But to do this systematically and in each separate departments a gigantic task. Not only is the domain to be mastered almost boundless; natural science in this entire domain is itself undergoing such a mighty process of being revolutionised that even people who can devote the whole of their spare time to it can hardly keep pace. Since Karl Marx's death, however, my time has been requisitioned for more urgent duties, and I have therefore been compelled to lay aside my work. For the present, I must content myself with the indications given in this book, and must wait to find some later opportunity to put together and publish the results which I have arrived at, perhaps in conjunction with the extremely important mathematical manuscripts left by Marx[9].

*** *** ***

London, September 23, 1885

[From: Engels, F. *Anti-Dühring* (Eng. Ed.), Progress, M., 1978, pp. 15-16, 18]

From Paul Lafargue's Reminiscences Of Karl Marx

*** *** ***

Apart from studying the works of poets and novelists, the other astonishing means that Marx discovered for giving rest to his brain was studying mathematics. He had a special kind of inclination to mathematics. Algebra even provided him with mental solace, he used to study this branch of mathematics during the most painful moments of his eventful life. He could not concentrate upon his usual scientific activities during the last days of his wife's illness, in those days he could forget the pain that seared his mind owing to her illness, only by studying mathematics. During these days of mental agony he wrote a treatise on infinitesimal calculus. Experts are of the opinion that this work of him is of immense scientific value. In higher mathematics Marx saw the most consistent and at once the most simple expression of dialectical movements. He was of the opinion, that so long as a science does not get used to the use of mathematics, it can not be called a truly mature science.

*** *** ***

[First published in 1890]

[*Marx-Engels Smriti*, Pragati, Moscow, 1976, pp. 31-32 (in Bengali)]
<An alternative translation of the passage:

Besides the poets and novelists, Marx had another remarkable way of relaxing intellectually – mathematics, for which he had a special liking. Algebra even brought him moral consolation and he took refuge in the most distressing moments of his eventful life. During his wife's last illness he was unable to devote himself to his usual scientific work and the only way in which he could shake off the oppression caused by her sufferings was to plunge into mathematics. During that time of moral suffering he wrote a work on infinitesimal calculus which, according to the opinion of experts, is of great scientific value and will be published in his collected works. He saw in higher mathematics the most logical and at the same time the simplest form of dialectical movement. He held the view that science is not really developed until it has learned to make use of mathematics.

URL:<https://www.marxists.org/archive/lafargue/1890/xx/marx.htm>

A note on the history of collecting, deciphering, editing and publication of Marx's mathematical manuscripts, 1994

Pradip Baksi

Karl Marx died in 1883. A partial edition of his *Mathematical Manuscripts* came out in print 85 years after his death, in 1968. During the first years of this intervening period, these manuscripts arrived in Germany, together with the other unpublished manuscripts of Marx and Engels, from England. These manuscripts became the property of the Social Democratic Party of Germany[10]. We know that Engels considered Marx's mathematical manuscripts to be important. He expressed his desire to publish them, in his preface to the second (1885) edition of *Anti-Dühring*. This wish remained unfulfilled during his life time. Frederick Engels died in 1895. After his death, it was the Social Democratic Party of Germany that became primarily responsible for publishing the said manuscripts. This party failed to fulfill this responsibility. What is more, one of the important leaders and theoreticians, first of the Social Democratic Party and subsequently of the Communist Party of Germany, Franz Mehring (1846-1919) declared, at the behest of some mathematicians (of whom exactly, we do not know), that Marx's mathematical manuscripts are of no importance [19, p. 14; all references in this article are to the entries in the Bibliography appended at the end of Part One, pp. 14-19]. Before the Russian revolution of 1917, a Russian revolutionary emigrant David Borisovich Riazanov (Goldendakh) (1870-1938) worked for some time in the archives of the Social Democratic Party, in Berlin. At that time he noticed that a part of Marx's mathematical manuscripts was not there in the archives. He located them at the residence of an important leader of the Social Democratic Party, Eduard Bernstein (1850-1932). Subsequently Riazanov approached a leader of the Austrian Social Democratic Party Frederick Adler (1879-1960) and, requested him to take the initiative for publishing the mathematical manuscripts of Marx. Riazanov's attempt too failed to bear any fruit, but in the process ten mathematical note-books of Marx went into the personal custody of Adler.

A few years after the revolution of 1917 a Marx-Engels Institute was established in Moscow and, Riazanov was appointed its first director. This institute of Moscow acquired a contractual right to photo-copy the manuscripts of Marx and Engels, from the archives of the Social Democratic Party of Germany. In pursuance of this contract Riazanov and his colleagues demanded the mathematical manuscripts of Marx for photo-copying purposes. It was only then that the authorities of the archives of the Social Democratic Party of Germany could recover the aforementioned ten note-books of Marx from Adler. Riazanov's efforts in this direction were reported in the July 1924 issue of "*Inprekorr*" published from Vienna [64]. At long last in 1925 the Marx-Engles Institute of Moscow succeeded in obtaining the photocopies of 865 pages of Marx's mathematical manuscripts [16, p. 56]. A German mathematician E. Gumbel was already acquainted with these manuscripts. He was brought to Moscow and given the task of editing them. R. Mateika and R.S. Bogdan helped him in the task of deciphering the texts. In 1927 Gumbel declared that the press-copy of Marx's mathematical manuscripts was ready [ibid]. However, some other associates of the Institute (for instance, E. Kolman) had a different opinion in this regard [see: 9, p. 185 and 29, p. 101]. Riazanov, the first director of the Institute, was expelled from the C.P.S.U in 1931. He was killed in 1938. Vladimir Viktorovich Adoratsky (1878-1945) became the next director of the Institute. Gumbel was removed and, in his place the task of editing the mathematical manuscripts of Marx was given to Sofya Aleksandrovna Yanovskaya (1896-1966). Initially she was assisted by D.A. Rykov and A.E. Nahimovskaya. In 1933, on the occasion of the 50th death anniversary of Karl Marx, two of his articles on the nature of differential calculus and, an editorial article of Yanovskaya were published in the journal "*Pod Znamenem Marksizma*" and in a collection of essays entitled "*Marksizm i estestvoznanie*". Upto 1968, those who were interested in the said mss had to remain contented mainly with these publications. After 1932, a Swedish mathematician Wildhaber remained associated, for sometime, with the Moscow-team working on the mathematical manuscripts of Marx. A member of the Soviet delegation to the Ninth International Congress of Mathematicians (1932) declared in one of the sessions of the congress that the entirety of Marx's mathematical writings are going to be published soon [27]. This promise too remained unfulfilled. During the Second world War the archives and the library of the Institute were shifted to some place in the Soviet Far East. Work of the Institute slowed down. After the War the pace of work picked up slowly. However, Yanovskaya—the editor of Marx's mathematical manuscripts—was also engaged with the teaching of mathematical logic in the Moscow University and, with the task of translating and editing text-books of mathematical logic. Her health began to deteriorate. A Congress of Mathematicians was organised in 1950 at Budapest. Here the delegates from some other communist-party-ruled countries repeatedly asked the members of the Soviet delegation: when, at long last, are they going to publish the mathematical manuscripts of Marx? The members of the Soviet delegation had no definite answer to this question [19, pp. 205-206]. After the return of this delegation to the USSR, the responsible authorities began to take more vigorous steps. Now Yanovskaya was given a new assistant: Konstantin Alekseevich Rybnikov. An important event of the 1950s was the publication of a note of Marx entitled "*On The Concept Of Function*" in the journal "*Voprosy Filosofii*" No. 11, 1958)[11]. As the work entered into the 1960s, it was noticed that the manuscripts are opaque in quite a few places (similar problems were being encountered by the editors of many other manuscripts of Marx and Engels). That is why the question of verifying the text once more from the originals was posed with some urgency. Meanwhile—in fact before the Second World War—on the 19th of May 1938, the entire archives of the Social Democratic Party of Germany, inclusive of the manuscripts and letters of Marx and Engels, has become the property of the International Institute of Social History, Amsterdam (see: note[10]). That is why, a Soviet delegation was sent to Amsterdam in August 1964, in search of the necessary papers [19, p. 207]. A member of this delegation was Olga Konstantinovna Senekina. She noticed, that when—during 1924-1930—the team of Soviet workers under the guidance of D.B. Riazanov, was busy in photo-copying some 55,000 pages of the writings of Marx and Engels, then quite a few pages of Marx's mathematical manuscripts remained un-photocopied, due to inadvertence. Now the photo-copies of those pages were obtained. This pushed up the volume of the hitherto known mathematical manuscripts of Marx to nearly 1,000 pages (see: the Preface to the 1968 edition of these mss). The editor of the 1968 edition of Marx's *Mathematical Manuscripts*, S.A. Yanovskaya died in 1966. Two years after her death, in 1968, a 639 page edition of Marx's mathematical manuscripts was brought out under the joint supervision of K.A. Rybnikov, O.K. Senekina and A. Ryvkin. In the Preface to this edition, the work that went on around these mss prior to 1931, remains unreported. However, the work of the 1933-1968 period has been described in detail. An even more important "silence" of this Preface involves the fact that it is an incomplete edition. This becomes evident to anyone who cares to go through this edition. Why were certain pages of Marx's mathematical manuscripts dropped? Apparently, the persons responsible for the publication considered these pages to be mathematically insignificant. It goes without saying, that such an approach towards the manuscripts of dead authors is—to say the least—unhistorical.

It may .be mentioned here that the work of the Institute of Marxism-Leninism of the CPSU (which evolved out of the Marx-Engels Institute of Lenin's time) has always been affected by the power-struggle within the top leadership of that party. Beginning with Riazanov, many competent workers were removed and killed. Those who were spared the hospitality of labour camps or death, left or were forced to leave the country. The noble work of editing and publishing the works of Marx and Engels became the subject-matter of self-promotion and intrigues of mean-minded and incompetent persons. In an atmosphere of all-round decadence of Soviet barrack socialism, quite a few generations of that society lost all interest in Marx, thanks to the criminal activities of the party and state leaders. Those who retained some honest interest failed to remain in the good books of the powers that be. In these circumstances, the 1968 partial edition of Marx's mathematical manuscripts was left just like that for about two decades. Even more truncated editions were brought out in some other countries (see: the bibliography). Words went around that the 1968 edition is by and large satisfactory and, that it would be included in that form in the contemplated complete works of Marx and Engels*. This "consensus" was broken in the wake of Perestroika. A decision was taken towards the end of 1987, to prepare a new and complete edition of Marx's mathematical manuscripts on the basis of all the hitherto available papers. At this stage, the task of editing was given to Irina Konstantinovna Antonova of the Institute of Marxism-Leninism, Moscow. Towards the middle of 1988 she told the present author that she hopes to complete her part of the job by the end of 1990. Then it would require the approval of the experts of the Institutes of Marxism-Leninism of Berlin and Moscow, before publication. These plans have been overtaken by larger movements of history. Tumultuous changes have taken place by the end of 1991. First the G.D.R. and then the U.S.S.R. has ceased to exist. The ruling communist parties of these two countries have been abolished

* For a different opinion on this question see: translation of Malodshii, 1983 in Part Two of this collection.

and the Institutes of Marxism-Leninism controlled by them have been wound up. The publication of a complete edition of Marx's mathematical manuscripts as part of the projected complete works of Marx and Engels has now become a matter of uncertainly all over again. When, where and how they would be published in future or whether they would be at all published ever—who knows?

NOTES TO REFERENCES IN THE PRECEDING PAGES

1. In the summer of 1857 Marx began to write a series of economic manuscripts in order to sum up and systematize the results of his extensive economic research started in the 1840s and continued most intensively in the 1850s. (In the first half of 1850s he filled 24 paginated and several unpaginated notebooks with excerpts from the works of other economists, books of statistics, documents and periodicals.) These manuscripts grew into: *Marx-Engels-Gesamtausgabe*, ZWEITE ABTEILUNG: "DAS KAPITAL" UND VORARBEITEN; URL: http://mega.bbaw.de/struktur/abteilung_ii.
2. Lange, Friedrich Albert (1828-1875): German philosopher, economist, neo-Kantian; member of the Standing Committee of the General Association of German Workers (1864-66), member of the First International, delegate to its Lausanne Congress (1867).
3. I began my inquiries about Hegel's mathematical manuscripts in 1980. In this connection, I received a letter from Dr. Helmut Schneider of the Hegel-Archiv, Ruhr-Universität Bochum. In this letter dated the 7th of February 1982, he stated that the mathematical manuscripts of Hegel are neither there in their archives, nor have these mss been published so far.—P.B.
   <Some of the related publications include:
   G.W.F. Hegel (1936), Geometrische Studien, in: *Dokumente zu Hegels Entwicklung*, S. 288-300; Hrsg. von J. Hoffmeister. Stuttgart: Frommann. 1936.
   __ (2008), "Geometrical Studies", *Hegel Bulletin*, Volume 29, Issue 1-2 (Number 57/58), 2008: 132-153.
   __ (2014), Geometrische Studien: Zwei Fragmente; in: *Gesammelte Werke*, I/2: Frühe Schriften, Teil II: 369-376; Bearbeitet von Friedhelm Nicolin, Ingo Rill und Peter Kriegel und Herausgegeben von Walter Jaeschke; Hamburg: Felix Meiner Verlag, 2014. See: Present Volume, Appendix II.>
4. *Moore, Samuel* (1838-1911): English lawyer, member of the First International, translated into English Volume One of *Capital* (in collaboration with Edward Aveling) and the *Manifesto of the Communist Party*; friend of Marx and Engels.
5. In this connection see:
   a) Zoltan, K. Marx és a válságok törtvényeinek matematikai tanulmányozása [Marx and the mathematical investigations into the

laws of crisis], *Közgazdasági Szemle*, Budapest, 1962, 12, pp. 1464-1483;

b) *Dunayaeva*, V. K voprosu O matematicheskom metode v "Kapitale" K. Marksa [On the question of mathematical method in Karl Marx's "Capital"], *Voprosy Ekonomiki*, 1967, 8, pp. 18-30.

6. "Moor" was Marx's nickname in Marx family. His close friends also called him by that name [see: Marx-Engels Reminiscences. Bengali ed. Progress. M., 1976].
7. See: Marx's manuscripts titled "On the Concept of the Derived Function" and "On the Differential"; in: Karl Marx, *Mathematical Manuscripts*, Delhi: Aakar Books, 2018.
8. See: Hegel, G.W.F., *Science of Logic* (Tr.: W.H. Jhonston and L.G. Struthers), G. Allen and Unwin, London, 1929; Volume One, Book One, Section Two, IIC—The Quantitative Infinity, pp. 241-332 [especially: IIC (c)—The Purpose Of The Differential Calculus Deduced From Its Application, pp. 291-320].
9. Here Engels expressed his desire to publish his *Dialectics of Nature* and Marx's *Mathematical Manuscripts* together. This wish of his remained unfulfilled. The Dialectics Of Nature was first published in 1925 and a part of Marx's *Mathematical Manuscripts* in 1933, an enlarged but nevertheless incomplete edition of the same came out in 1968.
10. For a general overview of the history of preservation, change of ownership and publication of the manuscripts, letters etc. of Marx and Engels, see: Saha, Dr. Panchanan—*Marx Engelser Pandulipi Kibhabe Raksha Pelo*? Indo-GDR Friendship Society, Calcutta, 1983. For a contemporary understanding of the state of Marx/Engels Papers go to: https://search.socialhistory.org/Record/ARCH00860
11. See: O Ponyatii Funktsii, *Voprosy Filosofii*, 1958, 11, pp. 89-95.

Bibliography

Different Editions of the Mathematical Manuscripts of Karl Marx

__ (1968), *Matematicheskie rukopisi*, Ed. Sofya Aleksandrovna Yanovskaya; Moskva: Nauka.

__ (1971), *Makesi shuxue shougao*; Shanghai: Fudan University, Science Section.

__ (1972), Reprint of *Makesi shuxue shougao* (1971); Nanjing: Teachers College.

__ (1974), *Mathematische Manuskripte*, Ed. Wolfgang Endemann; Kronberg/Taunus: Scriptor Verlag.

__ (1975), *Makesi shuxue shougao*; Beijing: Renmin chubanshe.

__ (1975), *Manoscritti matematici*, Tr. and Ed. Francesco Matarrese and Augusto Ponzio; Bari: Dedalo Libri.

__ (1983), *Mathematical Manuscripts of Karl Marx*; London: New Park Publications.

__ (1985), *Les manuscrits mathématiques de Marx*, éd. Alain Alcouffe; Paris: Economica.

__ (1990), *Matematiksel El Yazmalari*, Turkish translation of Karl Marx (1983) indicated above by Öner Ünalan; Demirtepe/Ankara: Başak Yayınlar.

__ (1994a), *Gonit Bishayak Pandulipisamuha, o bishesh songjojani: Marx o Gonit*; Bengali translation of: K. Marks (1968), *Matematicheskie Rukopisi* (Moskva: Nauka) together with a Special Supplement on *Marx and Mathematics*, Tr. and Ed. Pradip Baksi; Kolkata: Viswakos.

__ (1994b), *Mathematical Manuscripts*, together with a Special Supplement titled *Marx and Mathematics*; English translation of K. Marks (1968), *Matematicheskie Rukopisi* (Moskva: Nauka) with a Supplement, Tr. and Ed. Pradip Baksi; Calcutta: Viswakos.

__ (2005), *Manoscritti matematici*, Ed. Augusto Ponzio; Milan: Spirali.

__ (2018), *Mathematical Manuscripts*; second revised edition of the English translation indicated in (1994b), Tr. and Ed. Pradip Baksi; Delhi: Aakar Books.

Books and Articles on Marx's *Mathematical Manuscripts*

1. *Alcouffe A.*, Marx, Hegel, et le "Calcul" // *Marx, K*, Les manuscrits mathématiques, Paris: Economica, 1985. pp. 9-109.
2. *Baksi P.*, Markser Gonit Vishayak Pandulipi Prasange (On Marx's mathematical manuscripts) // *Gonit*, Calcutta, 1983, 2(2), pp. 51-58.
3. *Baksi P.*, Karl Markser Gonit Vishayak Pandulipi O TarTatparya (The mathematical manuscripts of Karl Marx and their significance) // *Marksbad, Gonit O Tarkashastra*—a collection of essays, Calcutta: Jnananveshan, 1985, pp. 30-65.
4. *Baksi P.* On the Problem of Situating Marx's Mathematical Manuscripts in the History of Ideas // Present Volume, Part Two.
5. *Basheleishvili B.*, K Voprosu o zakone edinstva protivopolozhnostiei v matematicheskikh rukopisyakh K. Marksa (On the question of the law of unity of opposites in the mathematical manuscripts of K. Marx) // Studencheskaya Nauchnaya Konferentsia, Tbilisi, 17-23 aprelia, 1952. Izd. Tbilisskovo gos universiteta im I. V. Stalina, 1952.
6. *Bottazzini U.*, Processi algebrici e processi dialettici nei Manoscritti Matematici di Marx, 1975. Unpublished. A later French version of it was published in: *Dijalektika*, Beograd, 1980, 3/4, pp. 69-85.

7. *Budach L.*, Karl Marx und die Matematikeinmal anders gesehen // Karl Marx, die Berliner Universität und die Verantwortung für die Festigung des Friedens und des Sozialismus, Berlin 1983, s. 34-40.
8. *Burkhardt F.*, Marx und die Mathematik // Wiss. Zeitschr. der Humbold.-Univ. zu Berlin, Ges.- u. Sprachwiss, Reihe, 1968, Hft. 6, s. 863-866.
9. *Burkhardt F.*, Karl Marx und die Mathematik // Karl Marx "Das Kapital". Erbe und Verpflichtung. Leipzig, Karl Marx-Universität, 1968, s. 695-701.
10. *Casanova G.*, Karl Marx et les mathématiques//*La Pensée*, n. s. Paris, 1948, N. 20. pp. 68-72.
11. *Endemann W.*, Einleitung // *Marx, K.*, Mathematische Manuscripte, Kronberg ts. Scriptor Verlag, 1974. s, 15-49.
12. *Gerdes P.*, Marx demystifies calculus // Studies in Marxism, Vol. 16, 1985; MEP Pubis. Minneapolis, USA. Reviewed in: Science and Nature, 7/8. pp. 119-123.
13. *Glivenko V.I.*, Poniatie Differentsiala u Marksa i Adamara (The Concept of Differential according to Marx and Hadamard) // *Pod Znamenem Marksizma*, 1934, 5, str. 79-85; and, present volume: 37-45.
14. *Gokieli L.P.*, Matematicheskie Rukopisi Karla Marksa i voprosy obosnovanya Matematiki (Mathematical Manuscripts of Karl Marx and the Questions of the Foundations of Mathematics). Izd AN Gruzinskoi SSR Tibilisi, 1947. (A Georgian edition of the book was published in the same year.)
15. *Guerraggio A.*, *Vidoni F.*, Nel laboratorio di Marx: Scienze naturali e matematika, Milano: Angeli, 1982.
16. *Gumbel, Emil Julius* (1927), "O matematicheskikh rukopisiyakh K. Marksa (Soobschenie)" ["On the mathematical manuscripts of K. Marx (A Communication)"], *Letopisi marksizma* [*Chronicles of Marxism*], 3/1927: 56-60. English translation in: Marx, 2018: ix-xiv.
17. *Herzmann J.*, *Tomek I.*, Marksova kritika metafyzickébo pojeti Základnich pojmu diferenciálniho poctu // *Filoz. Cas.*, Praha, 1981, 1, s. 78-93.
18. *Ibragimov S.*, Matematicheskie proizvedenya K. Marksa (Mathematical works of K. Marks) // *Azerbaijan Kommunist*, 1968, 1, str. 9-14.
19. *Katolin L.*, "Mee byli togda derzhkimi parnyami..." ("We were then the impudent twin.."). M. Izd. "Znanie", 1973 (2-oe izd. 1979).
20. *Kennedy H.C.*, Review of Karl Marx's *Matematische Manuscripte* (Scripter Verlag, F.R.G., 1974) and of *Manoscritti matematici di K. Marx* (Bari: Dedalo Libri, 1975) // *Historia Mathematica*, 1976, 3, pp. 490-494.
21. *Kennedy H.C.*, Karl Marx and the foundations of differential calculus // *Historia Mathematica*, 1977, 4, pp. 303-318; correction in: *Historia Mathematica*, 1977, 5, p. 92.
22. *Kennedy H.C.*, Marx, Peano and the Differentials. Unpublished. Revised copy of a paper read at the 15th International Congress of the History of Science, Edinburg, August 1977.

23. *Kennedy H.C.*, Marx's Mathematical Manuscripts // *Science and Nature*, 1978, 1, pp. 59-62.

24. *Kholshevnikov A.*, O matematicheskikh rukopisyakh Marksa (On Marx's mathematical manuscripts) // *Front nauki i tekhniki*, 1933, 2, str. 100-106.

25. *Kisileva N.A.*, Karl Marks i Matematika // *Matematika v shkole*, 1969, 1, sir. 5-10.

26. *Kolman A.*, Karl Marks i matematika // *Dijalektika*, Beograd, 1968, 3, s. 27-41.

27. *Kolman E.*, Eine neue Grundlegung der Differentialrechnung durch Karl Marx // Verhandlungen des Internationalen Mathematiker-Kongress, Zurich, 1932; II Band / Sektions-Vortrage, 1932, s. 349-351.

28. *Kolman E.*, Eine neue Grundlegung der Differentialrechnung durch Karl Marx // *Archeion. Archivio di storia della scienza*, 1933, 15, pp. 379-384.

29. *Kolman E.*, K. Marks i matematika (O 'Matematicheskikh Rukopisyakh' K. Marksa) // *Voprosy istorii estestvoznania i tekhniki*, 1968, 25, str. 101-112. For an English translation of the same see: *Mathematical Manuscripts of Karl Marx*, London, 1983: 217-234.

30. *Kondakov N I.* (Ed.), Matematicheskie rukopisi Karla Marksa // Logicheskii Slovar-Spravochnik. M., Nauka. 1975, str. 341.

31. *Kuzicheva Z. A.*, *Rybnikov K. A.*, Matematika v nauchnykh issledovanyakh K. Marksa (Mathematics in the scientific investigations of K. Marx) // Istoriya i metodologiya estestvennykh nauk, Vyp. XXXII, Matematika, Mekhanika. Izd. Moskovskovo Un-ta, Avgust, 1986, str. 3-13.

32. *Loi M.*, Review of: Marx, K., *Les manuscrits mathématiques* (Paris: Economica, 1985) // *La Pensée*, Paris, 1987. 256, pp. 129-131.

33. *Lombardo-Radice, L.*, Dai 'Manoscritti matematici' di K. Marx // *Critica Marxista-Quaderni*, 1972, 6. pp. 273-277.

34. *Magani L.*, Marx matematico: Il fondamenti del calcolo differenziale // Conoscere Marx. Milano: Angeli, 1983, pp. 55-89.

35. *Maiburova D.*, O matematicheskikh rukopisyakh K. Marksa // *Obshestvennye Nauki v Uzbekistane*, 1968,5, str, 26-30.

36. *Miller M.*, Karl Marx' Begrungung der differntialrechnung // Wiss. Z. Hochsch Verkehrswesen "Friedrich List". Dresden 16, DDR., 1969, s. 649-659.

37. *Molodshii V. N.*, O Matematicheskikh rukopisyakh K. Marksa // *Matematika v shkole*, 1969, 1, str. 10-23.

38. *Molodshii V. N.*, "Matematicheskie Rukopisi" K. Marksa i razvitie istorii matematiki v SSSR (Marx's "Mathematical Manuscripts" and the development of History of Mathematics in the USSR) // *Voprosy Istorii Estestvoznania i Tekhniki*, 1983, 2, str. 29-34; English translation in Part Two of the present collection.

39. *Mueller J.*, Karl Marx und die mathematik // *Tägliche Rundschau*, jg. 9, Nr. 72 (1953-54).

40. *Nahimovskaya A.*, Ab matematychnykh pratsakh K. Marksa // *Zapiski Bellaruskai Akademii Nauk*, kn. 1. 1933, str. 1-30.
41. *Novak I.P.*, O matematicheskikh rukopisyakh K. Marksa // Problemy Filosofii. Alma-Ata, 1968, str. 311-320.
42. *Panfilov V.A.*, Vliyanie materialisticheskoi dialektiki konechnovo i beskonechnovo na sposob differentsirovaniya K. Marksa (Influence of the materialist dialectics of the finite and the infinite on K. Marx's mode of differentiation). Dissertation. Dnepropetrovsk Gosudarstvenny Universitet, 1978. A copy of the mss exists in the INION Library, Moscow.
43. *Przelaskowski W.*, Prace matematyczne Karloa Marksa // *Economista*, Warszawa, 1960, 1, s. 159-165.
44. *Przelaskowski W.*, Pojecie funkeji w pracach Karloa Marksa // *Studia ekonomiczne*, Warszawa, 1960, 4, s. 187-203.
45. *Przhesmitsky V. I.*, Operativnyi logicheskii apparat rabotayuschii v "Kapitale" i "Matematicheskikh Rukopisyakh" K. Marksa (On the Operational Logical Apparatus operative in Karl Marx's "Capital" and "Mathematical Manuscripts") // Voprosy dialektichcskoi logiki, printsipi i formy myshleniya. M. 1985. s. 70-81; English translation in Part Two of the present collection.
46. *Psheliaskovsky V.*, Matematicheskie metody v ekonomii v svete "Matematicheskikh Rukopisiei" Karla Marksa (Mathematical Methods in Economics in the light of the "Mathematical Manuscripts" of Karl Marx). Doklad na mezhdunar. ekon. konf. 5-8 Sent., 1972, Varshva.
47. *Piaskovskii B.V.*, *Konvai B.I.*, K. Marks i matematika // Filosofski problemi suchasnovo prirodoznavstva. Mitvid. Naukovi zbirnik. Vip. 11, Kiiv, Vig. Kiivskovo Un-tu, 1968, str. 47-60.
48. *Rabinovich I.M.*, O poliarnom svoistve znaka ravenstva v "Matematicheskikh Rukopisyakh" K. Marksa (On the polar properties of the equality sign in K. Marx's "Mathematical Manuscripts") // Nauka i tckhnika, voprosy istorii i teorii. M.-L., 1977.
49. *Reiske G.*, *Sehenk G.*, Marx und die mathematik // *Deutsche Zeitschrift fur Philosophie*, Berlin, 1972, 4, s. 475-483.
50. *Ruzavin G.I.*, "Matematicheskie Rukopisi" K. Marksa i nekotorye problemy metodiki matematiki ("Mathematical Manuscripts" of K. Marx and some problems of the methodology of mathematics) // *Voprosy Filosofii*, 1968, 12, str, 59-70.
51. *Rybnikov K.A.*, O rabotakh K. Marksa po matematike (On K. Marx's works on Mathematics). Dissertation. 1954. Unpublished.
52. Rybnikov K.A., Matematicheskie Rukopisi Marksa // *Uspekhi Matematicheskikh Nauk*, 1955, 10(1), str. 197-199.
53. *Rybnikov K.A.*, K voprosu o poniatii funktsii (On the question of the concept of function) // *Voprosy Filosofii*, 1958, 11, str. 89-92.
54. *Ryvkin A.*, Marx y las matemáticas // *Documentos Politicos*, Bogota, 1968, N. 75. Mayo-Jun, pp. 34-38.

55. *Shugain O.V.*, Mathematichni rukopisi K. Marksa i "Kapital"// *Visnik Kiivskovo Universitetu*, 1968, Seria Filosofii, 2, str. 92-98.
56. *Slavkov S.*, Samostoiatclnite raboti na Karl Marks v oblasta na matematikata // *Spisanie na Bulgarsk. Akad. na Naukite*, Sofia, 1961, 4, str. 43-61.
57. *Slavkov S.*, Karl Marks i niakoi problemi na matematikata. Sofia, "Nauka i Izkustvo", 1963, 122s.
58. *Slavkov S.*, Karl Marks i niakoi voprosi na dialektikata v matematikata // Karl Marks i Filosofskovo Znanie. Sofia. Izd na BAN, 1984, s. 7-42.
59. *Smith C.*, Hegel, Marx and calculus // Mathematical Manuscripts of Karl Marx. London, 1983, pp. 256-270.
60. *Snij Mo-fu*, On the mathematical manuscripts of Marx (in Chinese). Sin kiesjuz, Peking, 1952.
61. *Struik D. J.*, Marx and Mathematics // *Science and Society*, 1948, 12, pp. 181-196.
62. *Toms M.*, Nad matematickými rykopisy Karla Marxe // *Politekonomie*, Praha, 1980, 6, s. 569-576.
63. *Treder H.-J.*, Die Beziehungen von Marx und Engels zu Mathematik und Naturwissenschaft // *Marx-Engels-Jahrbuch*. Berlin: Dietz, 1985, 9, s, 34-58.
64. *Varjas A.*, Marx als mathematiker. Ein Beitrag zur Entwiklung der Marx'schen dialektik // *Internationale Presse-Korrespondenz*, Wien, 1924, 18. VII. No. 92, s. 1165-1167.
65. *Yanovskaya S.A.*, O matematicheskikh rukopisyakh K. Marksa // *Pod Znamenem Marksizma*, 1933, 1, str, 74-115.
66. *Yanovskaya S.A.*, Predislovie (Preface), Matematicheskie Rukopisi K. Marksa. M., Izd. Nauka, 1968 (pp. 1-15 of the present volume).
67. *Zeman J., Zapletal I.*, Matematicke rukopisy K. Marxe a jejich Vztach k dnesku // *Filoz. Cos.*, Praha, 1983, 4, s. 536- 547.
68. *Zoltan K.*, Marx és a válságok törtvényeinek matematikai tanulmányozása (Marx and the mathematical investigations into the laws of crisis) // *Közgazdasági Szemle*, Budapest, 1962, 12, pp. 1464-1483.

Additions to the Bibliography

Alcouffe, Alain (1987), "Le calcul différentiel, les mathématiciens et les économistes au XIX° – K. Marx et H. Laurent, lecteurs de J.L. Boucharlat" ["The differential calculus,mathematicians and economists in the nineteenth century: K. Marx and H. Laurent, readers of J.L. Boucharlat"], *Sciences et techniques en perspectives*, 1987, 13: 47-69; URL: <https://hal.archives-ouvertes.fr/hal-01312597/document>

English translation: in part two of the present collection.

Alcouffe, Alain and Julian Wells (2009), Marx, maths, and *MEGA*^{2}; Paper presented at the 13th Annual Conference of the European Society for the History of Economic

Thought, University of Macedonia, Thessaloniki, Greece. URL: <https://mpra.ub.uni-muenchen.de/80535/1/MPRA_paper_80535.pdf>

Antonova, Irina Konstantinovna (1999), "Mathematische Manuskripte von Marx. Der Versuch einer kritischen Analyse" ["The Mathematical Manuscripts of Marx: Attempt at a critical analysis"], *Vestnik MGU, Reihe Philosophie* [*Newsletter of the Moscow State University, Series Philosophy*], No. 1, 1999: 46-62.

____ (2004), *Marksizm vne politiki* [*Marxism beyond politics*], Glava [Chapter] 3: Rekonstruktsia protsessa poznaniya na osnove matematicheskikh rukopisiei Marksa [Reconstruction of the process of cognition on the basis of the mathematical manuscripts of Marx]: 95-136. Moskva: Editorial URSS.

Batur, Bahadir *et al* (2018), "Marx'in Matematiksel El Yazmalari" ["Marx's Mathematical Manuscripts"], *Madde, Dialektik ve Toplum* (Ankara), Kasim 2018, Cilt 1, Siyai 4: 330-338.

Carchedi, Guglielmo (2008a), "Logic and Dialectics in Social Science, Part II: Dialectics, Formal Logic and Mathematics", *Critical Sociology,* Volume 34, Issue 5 (September 2008): 631-656.

____ (2008b), "Dialectics and Temporality in Marx's Mathematical Manuscripts", *Science & Society,* Volume 72, No. 4, (October 2008): 415-426.

____ (2011), Marx's Mathematical Manuscripts, Appendix Three of: *Behind the Crisis: Marx's Dialectics of Value and Knowledge:* 279-290, Leiden/Boston: Brill. It is a modified version of Carchedi, 2008a and 2008b above.

____ (2012), "Mathematics and Dialectics in Marx: A Reply", *Science & Society,* Volume 76, No. 4 (October 2012): 546-549.

Damsma, D. F. (2015), *On the articulation of systematic-dialectical methodology and mathematics*. Dissertation. Universiteit van Amsterdam. URL: <https://pure.uva.nl/ws/files/2326599/154609_DAMSMA_thesis_met_cover.pdf>

Dauben, Joseph Warren (1998), "Marx, Mao and Mathematics: The Politics of Infinitesimals", *Documenta Mathematica*, Extra Volume ICM III, 1998:799-809. URL: <http://www.emis.ams.org/journals/DMJDMV/xvol-icm/19/Dauben.MAN.html>

____ (2004), "Mathematics and Ideology: The Politics of Infinitesimals", *LlUll,* Volume 27, No. 59: 291-355; URL: <https://dialnet.unirioja.es/servlet/articulo?codigo=2040428>

Fahey, Charles, C.T. Lenard, T.M. Mills and Lex Milne (2009), "Calculus: A Marxist Approach", *Gazette of the Australian Mathematical Society,* Volume 36, Number 4 (September 2009), Technical Paper: 258-265. URL: <http://www.austms.org.au/Publ/Gazette/2009/Sep09/TechPaperFahey.pdf>

Gerdes, Paulus (2014), *The philosophic-mathematical manuscripts of Karl Marx on differential calculus: An introduction* (eBook). Maputo, Mozambique: the Author and Lulu.com.

Kusraev, Anatoly G. and Semen S. Kutateladze (1994), *Nonstandard Methods of Analysis*, Part I: Infinitesimal Analysis, Chapter I: Excursus into the history of mathematical analysis, 1.2 Karl Marx on Mysterious Differential Calculus. Dordrecht: Springer. It is an English translation of: *Nestandartnye metody analiza* (1990), Novosibirsk: Nauka, by L. Iouzina-Tarvainen.

Matthews, Peter Hans (2002), The Dialectics of Differentiation: Marx's Mathematical Manuscripts and Their Relation to His Economics. Middlebury College Economics Discussion Paper; URL: <http://sandcat.middlebury.edu/econ/repec/mdl/ancoec/0203.pdf>

Mobasher, Bamshad (1996), "Marx's *Mathematical Manuscripts:* A ReviewEssay", *Nature, Society, and Thought,* Volume 9, No. 1 (January 1996): 107-115. Review of: Karl Marx, *Mathematical Manuscripts* (1994), Calcutta: Viswakos.

Serman, Pascal (2013), "Karl Marx et le calcul infinitésimal" ["Karl Marx and infinitesimal calculus"], *Repères-IREM*. Num. 91. pp. 93-103; URL: <http://numerisation.irem.univ-mrs.fr/WR/IWR13008/IWR13008.pdf>

Smolinski, Leon (1973), "Karl Marx and Mathematical Economics", *Journal of Political Economy,* Volume 81, No. 5 (September-October 1973): 1189-1204.

Vogt, Annette (1995), "Emil Julius Gumbel (1891-1966): der erste Herausgeber der mathematischen Manuskripte von Karl Marx" ["Emil Julius Gumbel (1891-1966): The first editor of the mathematical manuscripts of Karl Marx"], *MEGA-Studien* (Amsterdam), 1995/2: 26-36. English translation: in part one of the present collection.

Vogt, Annette unter Mitarbeit von Erhard Scholz (2018), Artikel "Mathematische Manuskripte", in: Band 9/1, *Historisch-kritische Wörterbuch des Marxismus:* Spalten 331-341; Berlin: Argument Verlag.

Xu, Yibao (2013), Dialectics of Numbers: Marxism, Maoism, and the Calculus of Infinitesimals, Chapter 5: 165-196, in: *Mr. Science and Chairman Mao's Cultural Revolution: Science and Technology in Modern China,* Edited by Chunjuan Nancy Wei and Darryl E. Brock, Lanham etc.: Lexington Books.

Emil Julius Gumbel (1891-1966): the first editor of the mathematical manuscripts of Karl Marx

Annette Vogt

On Friday, the 9th of September 1932, a former subject of the Austro-Hungarian Empire, Arnošt (also called Ernst) Kolman (1892-1979)—who was then an employee of the Communist Academy founded in 1918 in Moscow, the "Capital of World Revolution"—delivered a lecture in Section VII, devoted to "Philosophy and History", of the 9th International Congress of Mathematicians held in Zurich. The lecture was on "A new foundation of differential calculus by Karl Marx".[1] He belonged to a delegation consisting of eleven scholars from the U.S.S.R., both as one of the four-member delegation of the "Government of U.S.S.R." and, also as a representative of the "Communist Academy (Moscow)".[2] He declared before the mathematicians present that:

"The Marx-Engels-Lenin Institute of Moscow possesses hundreds of photocopies of the so far unpublished manuscripts of Karl Marx in the fields of mathematics, natural sciences and technology. Within these manuscripts the works on mathematics come first. There are 31 different excerpts from the fields of arithmetic, algebra, analysis and geometry, 19 independent drafts, predominantly on the foundations of analysis, and also on the application of mathematics in political economy."[3] While referring to Marx's reflections on a history of the differential calculus, from within these papers, he stressed that Marx was a "true dialectician". He also announced that

1. E[rnst] Kolman, "Eine neue Grundlegung der Differentialrechnung durch Karl Marx" ["A new foundation of the differential calculus by Karl Marx"], in: *Verhandlungen des Internationalen Mathematiker-Kongresses Zürich 1932*, Bd. 2: *Sectionsvorträge* [*Proceedings of the International Congress of Mathematicians, Zurich 1932*, Vol. 2: *Section Lectures*] (Zürich, Leipzig, 1932), S. 349-351.
2. See, ibid. Bd. 1, S. 24, 54.
3. Kolman, "Eine neue Grundlegung der Differentialrechnung durch Karl Marx" ["A new foundation of the differential calculus by Karl Marx"], *loc. cit.* (Note 1), S. 349.

"Professor <Sofya Aleksandrovna> Yanovskaya is directing the publication of the entire mathematical writings of Marx" at an Institute of Moscow which is entrusted with editing these Works.[4] So far, and so incorrect.

This "old sheep of the Comintern flock" and a "Russian Czech", in the words of V. Lakšin, written in November 1963[5], had concealed only "a few things" in his usual way.

In 1931/32 the planned publication of Marx's mathematical manuscripts was still a part of a much more comprehensive project: that of the so-called "first" Marx-Engels-Gesamtausgabe (MEGA[1]). The work on these manuscripts had already started in the 1920s, at the Moscow Institute, led since its foundation in 1921 and till 1931, by the well-known scholar, one of the most important specialists in the history of scientific socialism and Marx-Engels-Research, David Borisovich Riazanov (Goldendach) (1870-1938).[6]

In 1925 the services of the German mathematician Emil Julius Gumbel were secured for the publication of the mathematical manuscripts of Marx. In June 1990 the present author found in Moscow some documents that clearly demonstrate the "editorial primacy" claims of Gumbel in preparing these manuscripts for printing.

Emil Julius Gumbel was born in Munich in 1891. There he studied mathematics and economics from 1910 to 1914, became an Assistant at the Department of Statistics and Actuarial Science in 1913, and received a Doctorate in 1914 with a work on population statistics. From 1915 to 1922 he studied physics, among other things, at the University of Berlin. He taught statistics at the University of Heidelberg from 1923. He gained greater fame through his publications, especially through his books against the Defence of the Realm, national socialists and "vigilante murderers". Already in the summer of 1932, he was fired by the University of Heidelberg, and emigrated to France. From there, to save himself, he relocated to the USA in 1940, where he died in New York in 1966.[7]

Owing to the mediations of Willi Münzenberg (1889-1940), Julius Schaxel (1887-1943)—a biologist from Jena, and Albert Einstein (1879-1955), the

4. Ibid. S. 351.

5. V. Lakšin, "'Novij mir' vo vremja Chruščeva (1961-1964). Dnevniki", *Znamja*, ["'Novy Mir' at the time of Khrushchev (1961-1964). Diaries", *Banner*,] 1990/7, S. 102.

6. This Institute was founded as the "Marx-Engels-Institute" at the beginning of 1921, and it was administered as a scientific facility since July 1922. In 1924 the XIIIth conference of the Russian Communist Party (Bolshevik) as well as the Vth World Congress of the Communist International decided that a complete edition of the Works of Marx and Engels will be published. Riazanov played a significant role in this decision-making process. He had also conducted the necessary negotiations with the Social Democratic Party of Germany in 1920 and 1924 and, had obtained their consent for photocopying and publishing the manuscripts, documents and letters <of Marx and Engels> held by them.

7. See: Annette Vogt, "Emil Julius Gumbel, Mathematiker und streitbare Publizist – auf der suche nach Wahrheit" ["Emil Julius Gumbel, mathematician and ardent publicist – in search of truth"], in: Emil Julius Gumbel, *Auf der Suche nach Wahrheit. Ausgewühlte Schriften*, hrsg. [...] mit einem Essay von Annette Vogt [Emil Julius Gumbel, *In Search of Truth. Selected Writings*, ed. [...] with an essay by Annette Vogt] (Berlin, 1991), S. 7-45. See also: Christian Jansen, *Emil Julius Gumbel. Portrait eines Zivilisten* [*Emil Julius Gumbel. Portrait of a Civilian*] (Heidelberg, 1991).

representative of the VOKS[8] in Berlin A.A. Štange addressed a letter to the Chairperson of the VOKS Olga Davidovna Kameneva[9] and, conveyed Gumbel's desire to go to Moscow[10].The Moscow office of the VOKS proposed to ask Riazanov if there was work for Gumbel at his institute[11]. Riazanov wrote back to the VOKS on the 17th of July 1925:

"I received your letter of 02 July, concerning the possible trip of Gumbel to the USSR. *On my part, I agree to entrust Gumbel with the work on the manuscripts of Marx (concerning mathematics)* and to employ him generally for several months with other scientific work at our institute."[12]

The VOKS completed the formalities and, informed Gumbel that he should contact Riazanov at the end of July, at the Institute for Social Research, 17 Victoria Avenue, Frankfurt-am-Main.

Thus in the winter of 1925 Gumbel came to the Marx-Engels-Institute in Moscow, about which he wrote in a book published in 1927:

"The scientific substantiation of Marxism certainly has a central role. The Marx-Engels-Institute founded and directed by D.B. Riazanov serves this purpose. The institute has one of the finest and most complete libraries of socialist literature, socialist in the widest sense of the word. The bibliophiles are especially delighted by an excellent collection of the old Utopians, the philosophers by a vast Hegel- and Fichte-library, and the socialists by a complete collection of the works of Marx and Engels in all their editions.

The entire legacy of manuscripts of the two founders of scientific socialism is available in photocopies. There is also a collection of original manuscripts, posters, proclamations and pictures pertaining to the history of socialism. Together with the technical staff, about 100 people work in this Institute. This includes some well-paid (which is rare in Russia) Marx-specialists. It has been announced that they will publish a philologically and historically accurate collection of all, including the hitherto unpublished, Works of Marx and Engels in German, in 42 volumes. The Institute publishes Russian translations, such as a collection of the early French materialists, and two Russian periodicals. The first issue of an exemplary German magazine dedicated to the history of socialism has been published. I do not think that

8. VOKS is the Russian acronym for the Soviet All-Union Society for Cultural Relations with Foreign Countries [Всесою́зное о́бщество культу́рной свя́зи с заграницей (ВОКС)] that was established in Moscow in 1925.
9. Olga Davidovna Kameneva (b. 1883), Chairperson of the VOKS since 1929, was one of the younger sisters of Lev Davidovich <Bronstein> Trotsky (1879-1940) and, the first wife of Lev Borisovich Kamenev (1883-1936). She was arrested under false accusations, sentenced in the autumn of 1935 and, shot dead in 1941.
10. A.A. Štange to O.D. Kameneva, 02 July 1925, Российский центр хранения и изучения документов новейшей истории [Russian Centre for Storage and Study of the Documents of Recent History; henceforth: RC], 71/d/N/65, Bl. [Sheet; henceforth: S.] 1.
11. R. Weller, Moscow Office of the VOKS to A.A. Štange, 15 July 1925, Ibid, S. 2.
12. D.B. Riazanov to the VOKS, 17 July 1925, Ibid, S. 3 [my emphasis – A.V.]. In the original there is "использовать его", that is "use [...] him".

a more finely equipped and well-endowed institute is available for the scientifically educated socialists."[13]

Gumbel stayed in Moscow until early April 1926.[14] He also met Ernst Czóbel, a long-time associate of the institute, who had a prominent place <among the authors> in the collection published in honour of Riazanov, on his 60th birthday[15] in 1930. Judging by their located letters from the years 1926 to 1930, it may be said that they were already united in a friendly relationship.[16]

These letters do confirm that Gumbel's priority was the publication of the mathematical manuscripts of Marx![17] There can be no doubt that it was Gumbel who first looked through these mathematical manuscripts, compared the copies in Moscow with the original ones in the party archive of the SPD in Berlin, and he not only prepared his own article, but also made the manuscripts ready for their publication. On 4 September 1926, he wrote from his French holiday resort to Czóbel about a contribution to the *Marx-Engels Archiv*:

"The short article consisting of three parts, namely, [the][18] one that is already lying with you, on Marx's work on the differentiation of a product and a report on his other mathematical studies, has not appeared in the *Letopis* [*Annals*] for reasons unknown to me. I am currently working on this report, which will, of course, include numerous quotations, including one about an arc. The two first-mentioned parts are completely ready for printing."[19] The *Letopis* here refers to the journal *Letopisi marksizma* published by the Marx-Engels-Institute under Riazanov's leadership

13. E.J. Gumbel, *Vom Rußland der Gegenwart* [*From Russia Today*] (Berlin, 1927); cited from: Gumbel, *Auf der Suche nach Wahrheit* [*In Search of Truth*] (Note 7), pp. 83-164, p. 144. – Under Riazanov's direction the Institute published the Летописи марксизма, Записки Института К. Маркса и Ф. Энгельса (Москва, Ленинград 1926-1930) [*Annals of Marxism, Notes of the Institute of K. Marx and F. Engels* (Moscow, Leningrad 1926-1930)], Issues 1-11, the Бюллетень Института К. Маркса и Ф. Энгельса при ЦИК СССР [*Bulletin of the Institute of Karl Marx and F. Engels of the Central Executive Committee of the USSR*] (since 1926), as well as – since 1924 – the Архив К. Маркса и Ф. Энгельса [*Archives of K. Marx and F. Engels*] in Russian and – since 1926 – the *Marx-Engels-Archiv* in German. However, of the German edition, printed in Frankfurt/M. by the Leipzig publishing house J.B. Hirschfeld, there appeared only two volumes.

14. Gumbel wrote to Ernst Czóbel in Moscow on July 16, 1926: "If, as is unlikely, you were to come to me with a question (x) [enclosed in handwriting by the Ministry of Education in Karlsruhe], I beg you to confirm that I have finished my business in the Marx-Engels-Institute at the beginning of April. Yours cordially, Gumbel." RC 71/d/N/65, S. 6.

15. На боевом посту: сборник к шестидесятилетию Д. Б. Рязанова [*On the line of duty: a collection for the sixtieth birthday of D. B. Riazanov*] (Москва, 1930). This comprehensive anthology of 652 pages appeared in an edition of 3000 copies. Eleven months later, in early 1931, Riazanov was arrested.

16. According to the archive staff (as of June 1990) the correspondence of Riazanov did not survive. – Gumbel's letter to Riazanov dated October 18, 1930, may have been saved and found, because it was with other papers. After Riazanov's arrest, there was a "cleansing" at the institute, which killed many of the approximately 100 employees. – Ernst Czóbel (1886-1948) was also arrested, though he was innocent and, was sentenced to imprisonment in 1937.

17. With this statement the author would also like to rectify her own statement that appeared 13 years ago in: Annette Vogt, ""Karl Marx und die Mathematik" ["Karl Marx and Mathematics"], *Mitteilungen der Mathematischen Gesellschaft der DDR* [*Communications of the Mathematical Society of GDR*] [Berlin (DDR), 1983], H. 3, S. 50-61.

18. In the original: the inappropriate German article der [of the] has been written, in place of the appropriate article dem [the].

19. Gumbel to Ernst Czóbel, 4 September 1926, RC, 71/d/N/65, S. 10.

from 1926-1930. The publication of the "short article" mentioned by Gumbel, is his communication "On the mathematical manuscripts of K. Marx", reproduced below in the appendix, was postponed from Issue number 1 to Issue number 2 of the journal in 1926, and was finally moved to its Issue number 3 in 1927.[20] Gumbel pointed out that he had been commissioned by Riazanov to "look through the papers and prepare them for publication." He had decided "to bring all the materials into a systematic order and to prepare the relevant parts for printing." This task has presently been accomplished. The material was classified into: "A. Calculations, B. Excerpts, C. Drafts, and D. Independent Works". In his classification, he was guided by "purely formal considerations". After giving a detailed list of all the manuscripts (indicating the language in which they were written), Gumbel announced:

"His independent work on the differentiation of products is of greatest interest; it will be published in German in the next issue of the journal '*Archiv K. Marksa i F. Engelsa'*. All the works indicated through the numbers 1-13 above are ready for the press, and these will appear in Volume XVI of the Collected Works [i.e., of the *MEGA*[1]- *A.V.*]. There, the philosophical problems investigated by Marx in his mathematical works will come under critical reflection."

In September and October of 1926, Gumbel stayed in Berlin and compared the copies with the Marx's papers in the Archives of the Social Democratic Party of Germany.[21] He also prepared his contribution to the *Marx-Engels-Archiv*. However, by that time, in Moscow, the planned volume XVI of MEGA had already been postponed indefinitely. Initially Gumbel was not worried. On November 3, 1926, he wrote to Czóbel: "Your message that the 16th volume has been postponed is indeed a regrettable one, but I was glad that I could make it to the earlier publication of the first volumes."[22] However, the volume never came out.

The first volume of the *Marx-Engels-Archiv* was published in 1927. Gumbel's contribution was originally scheduled to be published in the second volume[23], which came out in the same year. Already on 13 October 1926 Gumbel had written from Berlin to Czóbel:

"I'll send you the rest of the comments at the same time. It contains everything philosophically important [...] Apart from your possible amendments to the comments, my whole work is now finished. It contains 69 pages, namely:

1) "The mathematical manuscripts of Marx and Engels" (6 pages) located by you,
2) the comments (33 pages),
3) the two remarks (2 pages),

20. Э. Гумбель, "О математических рукописях К. Маркса (Сообшение)", *Летописи марксизма*, 1927/3, стр. 56-60; published for the first time in German as: "Über die mathematischen Manuskripte von K. Marx (Mitteilungen), in: Gumbel, *Auf der Suche nach Wahrheit* (Note 7), S. 182-189. <A translation of this Communication from the Russian original is available in the Second and revised (Delhi, 2018) edition of the English translation of K. Маркс, *Математические рукописи* (Москва, 1968). –Tr.>
21. Gumbel to Czóbel, 29 September and 2 October 1926, RC/71/d/N/65, S. 12, 13.
22. Gumbel to Czóbel, 3 November 1926, ibid, S. 14.
23. Czóbel to Gumbel, 27 October 1926, ibid, S. 20.

4) the MS No. 4 located by you, together with my comments, including 28 pages. [...] It would be very important for me, if I could possibly get 100 offprints [...]."

He declared in this letter that he was ready to shorten his article, "if 69 pages are too much for an essay in the Archiv."[24] Already in February 1927 Gumbel got the first galley proofs.[25] However, in the second volume — Gumbel received this volume on 6 February 1928 — Friedrich Engels' "Dialectics of Nature" was published for the first time.

Finally, Gumbel's contribution to the third *Archiv*-Volume was scheduled.[26] However, in 1928, at first he had to make various changes in the introduction. On 10 October 1928 Czóbel finally wrote to him: "Unless the signs deceive, the third volume of the *Archiv* will now be put to serious work and you will soon hear from us again."[27] Apparently, Gumbel did not hear anything, because on 21 November he wrote — this time not to Czóbel, but officially to the institute — quite emphatically:

"Comrades, two years have passed since the dispatch of the manuscript titled 'The Mathematical Manuscripts of Marx and Engels'. From different sides, I have been asked quite bluntly, what had actually been the result of my Marx studies. It is very unpleasant for me to have to avoid such inquiries. I would be grateful if you could give me a definite answer about the date of publication of my article [...]."[28] Then, in December 1928, he received a letter of intent from Karl Schmückle.[29] However, on 21 March 1929 Czóbel informed him:

"In fact, we can not specify a fixed date for the publication of the 3rd volume of the *Archiv*. At the moment [...] when the final word falls — from whose mouth, you know very well — [...] the volume [can] come out in a few months. Unfortunately, the decisive word is not there. And I am really amazed that you, who know the situation in Moscow and in the Institute, cannot put yourself in our situation. The situation is now particularly complicated by the fact that on the one hand Ri[azanov] is sick, and on the other he is burdened with an immense amount of urgent work and tasks [...]. He fights to the right and to the left, reorganizes the Leningrad Academy. He has a great deal to do, and without him the volume cannot get out of the spot it is in. [...]

I request you to supplement my remarks with your memories of Moscow and some fantasy."[30]

It was the year 1929, which was later on called the year of "turnaround" (rupture). Obviously, the situation was such that Czóbel could only provide intimations or

24. Gumbel to Ernst Czóbel, 13 October 1926, ibid, S. 16 ff.
25. Marx-Engels-Institute (MEI) to Gumbel, ibid, S. 35. On 20 May 1927, Gumbel informed Moscow that the corrections have been completed, ibid, S. 53.
26. See Ernst Czóbel to Gumbel, 10 October 1928, ibid, S. 89; Karl Schmückle to Gumbel, 6 December 1928, ibid, P. 92. — <Karl Schmückle [1898-1938] was a German historian and literary scholar, a member of the KPD since 1919 and, that of the CPSU since 1926; he took part in editing four volumes of *MEGA*^{1}. > Schmückle too was one of the innocent convicts. He was shot dead in March 1938.
27. Czóbel to Gumbel, 10 October 1928, *loc. cit.* (Note 26).
28. Gumbel to the Marx-Engels-Institute, 21 November 1928, RC/71/d/N/65, S. 91.
29. Schmückle to Gumbel, 6 December 1928, ibid, S. 92.
30. Czóbel to Gumbel, 21 March 1929, ibid, S. 94.

suggestions. Just a few weeks after the arrival of this letter the April-Plenum of the Central Committee of the RCP (B) was held, in which there began the campaign against the "right opposition" i.e., that against Bukharin, Rykov and Tomsky.[31] At the November-Plenum of 1929, Nikolai Ivanovich Bukharin was removed from the Politburo. He was one of those who were closely related to Riazanov and his institute.

Still, both the sides did not give up the hope that Gumbel's work might appear. Finally, the third volume of the *Marx-Engels-Archiv* that was initially slated to be published in the autumn, appeared in the winter of 1930.[32] Gumbel had once again read the proofs in May 1930 and then heard nothing about them. So he wrote in person to Riazanov on 18 October 1930:

"Dear Comrade Riazanov,

You are in possession of my article titled 'The Mathematical Manuscripts of Marx and Engels' for the last 4 years. The final proofs are lying with you for the last 2 years. You will therefore find it understandable that I am interested in its early publication. I am fully aware of the great difficulties that you are facing. I have always taken them into account. However, now I ask for definitive information about when this article is going to get published.

With socialist greetings,

I remain

[Signed] Gumbel."[33]

Whereupon Riazanov replied on 25 October 1930:

"I can, however, reassure you that no one will obstruct the publication or handling of these studies of Marx [...]. This work will definitely get published in the next — third — German volume of the Archiv. This volume will certainly appear within this working year, by the summer of 1931. I can assure you that the institute, which got the editorial work [...] by you, has no less interest in the publication than you yourself have."[34]

Four months later, on 16 February 1931, Riazanov, who was known to be critical of Stalin, was arrested "by the Organs", as they were then called. He was extremely falsely accused in the so-called trial against the "counterrevolutionary

31. Nikolai Ivanovich Bukharin (1888-1938), Alexei Ivanovich Rykov (1881-1938), Chairman of the Council of People's Commissars 1924-1930, and Mikhail Pavlovich Tomsky (1880-1936), Chairman of the Central Trade Union Council since 1919, were regarded as the leaders of the so-called "right-wing opposition", after the inner party disputes of the previous years with the so-called "left-wing opposition" around Lev Davidovich Trotsky (Bronstein) (1879-1940), Grigory Yevseevich Zinoviev (Radomyslsky, Hirsch Apfelbaum) (1883-1936) and Lev Borisovich Kamenev (Rozenfeld) (1883-1936) were switched off. All of them were presented in the notorious show trials of the 1930s and then shot; Tomsky escaped from this fate by suicide, Trotsky was murdered in exile in Mexico. For the trials — illustrated by the example of <Martemyan Nikitich> Ryutin <1890-1937> — see Annette Vogt, "Eine bestechende Analyse, eine Fundierte Kritik, aber... — die Tragik des Martemjan Nikititsch Rjutin" ["A striking analysis, a profound criticism, but...— the tragedy of Martemyan Nikitich Ryutin"], in: Theodor Bergmann, Mario Keßler (Hrsg.), *Ketzer im Kommunismus — Alternativen zum Stalinismus* [*Heretics in Communism — Alternatives to Stalinism*] (Mainz 1993), S.140-161.

32. See the three letters to Gumbel dated 16 May, 28 May and 13 October 1930, RC, 71/d/N/65, S. 100-102.

33. Gumbel to Riazanov, 18 October 1930, ibid, S. 103.

34. Riazanov to Gumbel, 25 October 1930, ibid, S. 104. — This is the last letter in the folder, which the author got in the archive, for the sake of understanding.

organization of the Mensheviks" (1-9 March 1931).[35] He was banished to Saratov, but following an appeal from Leningrad's party secretary Sergei Mironovich Kirov (1886-1934), he was given an opportunity to stay in Leningrad. He had to return to Saratov after Kirov's assassination on 1 December 1934. There he was arrested again on 23 June 1937, sentenced to death under false charges on 21 January 1938 and, shot on the same day.[36]

After Riazanov's arrest, a "cleansing" began at the institute, and anyone who had worked with him was at least a "suspect". When Gumbel was again in Moscow in the winter of 1931-32, he was likely to have learned of the extent of the reprisals, as he had a command of the Russian language. However, he did not say anything about it in his travel notes titled "Moscow 1932" published in the journal *Weltbühne*.[37] At that time he could not have foreseen that only a few months later Kolman would speak of the intended publication of the manuscripts in Zurich, without wasting a word about Gumbel. Only a coincidence prevented him from being present there. He had participated in the International Congress of Mathematicians in Bologna in 1928, and then again attended the Congress in Oslo in 1936, but he could not go to Zurich because he just had to look for a new job. He had already been removed from the University of Heidelberg, following massive pressure from the national socialist students, even before the National Socialists came to power. So he could not comment on Kolman's presentation in Zurich.

Gumbel's work remained unpublished. Kolman, who left USSR in 1976, wrote in his memoirs that after the removal of Riazanov he was sent to his successor Vladimir Viktorovich Adoratsky (1878-1945). He had become a member of the Institute's board of directors and, the director of the Marx Cabinet, i.e. of the department responsible for the publication of the works of Marx. On the first day, the new directorate had approached the staff who had accused Riazanov of keeping Marx's manuscripts secret. Unfortunately, this is likely to be true. At that time it was a "normal" practice to charge others to save one's own skin, but it could hardly have been about some "secret" texts of Marx. According to Kolman,

35. See, Рой Медведев, "О Сталине и сталинизме" [Roy Medvedev, "On Stalin and Stalinism"], *Знамя*, 1989/2, 183, 188f.

36. Riazanov had never pleaded "guilty". He was rehabilitated in 1958. However, he was still an "Unperson" until the time of "Perestroika", when reports about him started coming out for the first time. See: В. А. Смирнова, "Первый директор Института К. Маркса и Ф. Энгельса – Д.Б. Рязанов," ["The First Director of the Institute of K. Marx and F. Engels – D.B. Riazanov"] *Вопрсии истории КПСС*, 1989/9, стр.71-84. In 1990, on the occasion of his 120th birthday, two volumes of his selected works were slated to appear. See also: Рокитянский Я. Г. "Неизвестные письма российских ученых двадцатых годов" ["Unknown letters of Russian scholars of the twenties"], *Вестник АН СССР*, 1991, № 11, ctr. 92–118; id. "Судебная расправа 1922 года: академик Рязанов против карательной практики большевиков" ["The judicial reprisal of 1922: academician Riazanov against the punitive practice of the Bolsheviks"], *Вестник РАН*, 1992, №4, стр.103–121; id. "Das tragische Schicksal von David Borisovič Riazanov" ["The tragic fate of David Borisovich Riazanov"], *Beiträge zur Marx-Engels-Forschung NF* 1993, S.3-16. Also see: *David Rjasanow: Marx-Engels-Forscher, Humanist, Dissident* [*David Riazanov: Marx-Engels-Researcher, Humanist, Dissident*], Hrsg. [...] von Volker Külow und André Laroslawsky. - Berlin: Dietz, 1993. <Also go to: http://rjazanov.shpl.ru/>.

37. See: E. J. Gumbel, "Moskau 1932", *Die Weltbühne*, 28 (1932), S. 400-403, 513-515, 591-593; reprinted in: Gumbel, *Auf der Suche nach Wahrheit* (Note 7), S. 165-178.

"a safe in Riazanov's former study had been opened with the help of a specialist [?!]"; it was found to contain letters from Marx and Engels "as well as Marx's mathematical notebooks and his notes on questions of the natural sciences".[38] "As far as the mathematical works of Marx are concerned, it seems to me from today's point of view, that Riazanov kept secret the work on deciphering and preparing for publication, because he was not convinced of their scientific value and also because he was afraid that they might damage the authority of Marx. The mediocre German mathematician Gumbel, to whom Riazanov gave them for assessment, did not understand their methodological importance and advised in favour of their publication [?!]."[39] Kolman then emphasized the merit of his own work on those manuscripts, and that of drawing Sofya Aleksandrovna Yanovskaya (1896-1966) and a certain <D.A.>Rykov into this work. However, because of another change in the Directorate — <after> 1938 Adoratsky was replaced <by M. B. Mitin> as its director — and the conditions prevailing in the 1930s, the publication was stopped. It was only after some 20 years that the careerist <K.A.> Rybnikov had all the glory and bagged a doctorate for this work that he had not done at all.[40]

It would be difficult to disprove Kolman's narrative without any knowledge of the above-indicated documents from Moscow. It may be surmised from his contemptuous remark about Gumbel that the two may have met in Moscow but had not come close to each other in person. When Kolman wrote his memoirs he could assume that since the short "Communication" of Gumbel in the *Letopisi marksizma* was difficult to access, hardly anyone would know about it. Like its editor, the journal was also neglected for decades in the Soviet Union. Moreover, others already had the fame of first discovery and publication of the manuscripts in question. When Kolman especially insults Konstantin Alekseevich Rybnikov, then it seems that there could be some envy involved in the game.

In fact, Marx's mathematical manuscripts were published in German and Russian in Moscow in 1968. There is no mention of Gumbel in its Preface, not even of Kolman.[41] At that time the latter had "difficulties" in connection with the events in Prague.

---

38. Арношт (Эрнест) Кольман, Мы не должны были так жить [*We should not have lived thus*] (New York, 1982), S. 171; German edition: Arnošt Kolman, *Die verrierte Generation. So hätten wir nicht leben sollen. Eine Biographie* [*The Lost Generation. We should not have lived thus. A Biography*] (Frankfurt/M., 1979), S. 139. The author has also used the German edition, but quoted from the Russian edition.

39. *Ibid.* S. 172 or, 139f.

40. *Ibid.* S. 172 or, 140. <Konstantin Alekseevich Rybnikov (1913-2004) defended his doctoral dissertation titled "Исследование математических рукописей Маркса" ["Investigations on the mathematical manuscripts of Marx"] at the Lomonosov Moscow State University on the 25th of June 1954; see: "Константин Алексеевич Рыбников. Творческий путь" ["Creative Path of Konstantin Alekseevich Rybnikov"], Чебышевский сборник [Chebyshev's Collection], T. 15, Вып. 4 (2014): 148-79, Стр.166. – Tr.>.

41. К. Маркс, *Математические рукописи*, Институт марксизма-ленинизма при ЦК КПСС (Москва, 1968). The Preface was written by S. A. Yanovskaya; A. Z. Ryvkin, < O. K. Senekina> and K. A. Rybnikov have been mentioned as collaborators. See: *ibid,* "Предисловие", S. 3-22, S. 22. – The following edition of Wolfgang Endemann, which appeared six years later, is based on this edition: Karl Marx, *Mathematische Manuskripte*, hrsg., eingel., und kommentiert von Wolfgang Endemann, Scriptor Taschenbücher, Sozialwissenschaften S 10 (Kronberg Ts., 1974).

Presumably, it was indeed the case that an anxiety that Marx's mathematical work might not match entirely up to the myth about his genius, first delayed its printing process, and then prevented it from being published for a long time. An excerpt from the expert's report by Gen. Thalheimer, which Gumbel sent with the letter dated 15 November 1926, suggests that the responsible authorities in Moscow were concerned about it. They were expecting such paeans from Gumbel, as were beyond him. The expert wrote, "Gumbel gave the material in an accurate form, ready for use; his edition of Marx's work is what a *Mathematician* can do about it: present it for the reader in an orderly form."[42] In April 1928 Gumbel received some "hints" about his essay, from Dr. <Aleksei Mikhailovich> Voden, who was a colleague of him in Moscow, including the following: "Your remark that: 'However (this comparison of the Taylor's Series with the Binomial Theorem) was not put forward by him, it is rather customary' is quite unnecessary. The Institute of Marx and Engels thinks that it would not be expedient to publish such remarks in print, so the best course for us would be to strike out this gloss."[43]

Not once did Gumbel settle for a discussion with Dr. Voden. He had already given his own assessment — his understanding, one may say, of Marx's mathematical works—in his "Communication" of 1927: Marx had mainly dealt with the works of those authors who were already known by the end of the 18th century. <After listing the mathematical texts studied by Marx, he wrote:> "It is clearly visible that his choice of these texts was influenced by that of Hegel." Hegel mentioned these very works of Euler, Lagrange, Landen and Newton and, he too was preoccupied with the foundations of differential calculus, and with the deduction of derivatives. The newer mathematical literature, such as the work of Gauss or Bolzano, was not known to Marx.

"The manuscripts of Marx that do not contain <mere> calculations or excerpts may be considered to be his *independent* work. Publishing only those make <some> sense. They contain a free exposition of what Marx studied, combined with numerous chronological dates, and philosophical reflections on what he read. In some cases Marx posed some problems in a special way that is characteristic only of him."

In certain cases the manuscripts so supplement each other that it becomes possible to "obtain a general picture of what Marx had in view."[44]

42. "Excerpt from a report by Gen. Thalheimer", RC, 71/d/N/65, S. 24 R. – Here "Gen. Thalheimer" might refer to <the German Marxist politician and theoretician> August Thalheimer (1884-1948), who lived in the Soviet Union <and worked for the Comintern and the Marx-Engels-Institute> from 1923 to 1928.

43. "Dr. Voden: On Dr. Gumbel's remarks on Marx's mathematical works", included in the letter to Gumbel dated 28 April 1928, RC, 71/d/N/65. S. 79. <On Алексей Михайлович Воден [Aleksei Mikhailovich Voden] (1870-1939?) see:
https://ru.wikipedia.org/wiki/%D0%92%D0%BE%D0%B4%D0%B5%D0%BD,_%D0%90%D0%BB%D0%B5%D0%BA%D1%81%D0%B5%D0%B9_%D0%9C%D0%B8%D1%85%D0%B0%D0%B9%D0%BB%D0%BE%D0%B2%D0%B8%D1%87>.

44. See Annex, S. 40. <This corresponds to: Эмиль Юлиус Гумбель, "О математических рукописях К. Маркса (Сообщение)," *Летописи марксизма*, 1927, 3, стр. 59, used for the present rendering. – Tr.>

Gumbel's friend Berthold Jacob (1898-1944) expressed himself much more sceptically — and rather drastically — in a review of Gumbel's book *Vom Rußland der Gegenwart* [*From Russia Today*]:

"Gumbel measured this Asiatic land with Karl Marx's yardstick, whose mathematically dilettante texts he was assigned to edit into a showpiece of Soviet-Marxist relic, in hundred-towered Moscow".[45]

However, even during the period of exile in France, when owing to the notorious Moscow trials an anti-Soviet mood increased among the German emigrants, Gumbel still acknowledged his work in Moscow. Thus he stated in the "Sammelbuch aus der deutschen Emigration" ["Collective Book from the German Emigration"], edited by him and published from Strasbourg in 1938, for which he wrote a small autobiography and a bibliography: "I worked at the Marx-Engels-Institute in Moscow during the winter vacations of 1925/26 and to make the mathematical notes left by Marx and Engels printable."[46]

When he wrote this, Riazanov had already been condemned to death as an "enemy" and a "counterrevolutionary", and Bukharin had to go through the show trials, which also went, inter alia, around Marx-Engels-Research,[47] and his friend from his days in Moscow Ernst Czóbel tried to survive through the GULAG-System, its pre-trial detention and "judgement"...

Karl Marx brought scant luck to those who dealt with his papers in Moscow in the 1920s.

<After this there is the:

Anhang: E. Gumbel, Über die mathematischen Manuskripte von K. Marx (Mitteilung). It is a German translation of Gumbel's Communication of 1927 published in Russian in the *Летописи марксизма* (see note 20 above). Its English

---

45. Barthold Jacob, *Vom Rußland der Gegenwart* [Rezension], *Die Weltbühne,* 23 (1927), 1. Hlbbd. S. 879. – See a totally different review of Gumbel's book by Peter Panter [i.e. Kurt Tucholsky], "Auf dem Nachttisch" ["On the Night Table"], *ibid,* 2. Hlbbd. S. 861 f.: "[…] the best that I have read about Russia lately is: E. J. Gumbel's 'From Russia Today' […]. Hundred pages – but that packs a punch. The best seems to me to be the passage where Gumbel speaks of the inky-black and chubby-pink descriptions that we get to read about Russia. […] However, Gumbel is not as enthusiastic as the overzealous traveller, whose Hi there stuff always seems a bit childish: the good Russians! […] They have the sun in their hearts…Gumbel has understanding". <See: http://www.textlog.de/tucholsky-nachttisch-molnar.html>

46. *"E. J. Gumbel" in: Freie Wissenschaft. Ein Sammelbuch aus der deutschen Emigration* [*Free Science. A Collective Book from the German Emigration*] hrsg. von E. J. Gumbel (Strasbourg, 1938), S. 267f. S. 268; reprinted in Gumbel, *Auf der Suche nach Wahrheit* (Note 7), S. 231-233.

47. See also, in particular, on the trials of Bukharin, accused for his travel to Paris in the spring of 1936 (to conduct negotiations on the Marx-Engels-Papers; so as to cultivate his connections with the "counterrevolutionaries" abroad), the memoirs of his widow Anna Mikhailovna Larina (1914-1996), who accompanied him: A.M. Ларина, "Незабываемое. Воспоминания", Знамя, 1988, H. 10, стр. 126-165; H. 11, стр. 112-180; H. 12, стр. 93-169; in German: Anna Larina Bucharina, *Nun bin ich schon bald über Zwanzig. Erinnerungen* (Göttingen, 1989) <English translation: Anna Larina, *This I Cannot Forget: The Memoirs of Nikolai Bukharin's Widow,* New York: W.W. Norton, 1994>.

translation from the Russian original is available in the second revised edition of Karl Marx, *Mathematical Manuscripts* (Delhi, 2018)>.

Source: "Emil Julius Gumbel (1891-1966): der erste Herausgeber der mathematischen Manuskripte von Karl Marx", *MEGA-Studien* (Amsterdam), 1995/2: 26-36.

Author: Professor Dr. Annette Vogt (*1952)
https://de.wikipedia.org/wiki/Annette_Vogt

Senior Research Scholar,
Max-Planck Institut für Wissenschaftsgeschichte
https://www.mpiwg-berlin.mpg.de/en/users/vogt

and, Honorary Professor,
Humboldt-Universität zu Berlin
https://www.wiwi.hu-berlin.de/de/administration/employee/1688215

Part Two

Investigations

Investigations Inspired by Marx's Mathematical Manuscripts: A Selection

The Concept of Differential according to Marx and Hadamard

Valery Ivanovich Glivenko

1. Two points of view on the concept of differential.

In the history of differential calculus one comes across two basic points of view on the concept of differential.

According to the first, the concept of differential *immediately* reflects some external reality; for the sake of brevity we shall call it the objective point of view. This was the point of view of the inventors of differential calculus. For them—the differential was an infinitesimal increment of the variable. The external reality reflected in the words "infinitesimal increment", was somehow thought to be self-evident.

The concept of derivative was not there, its role was fulfilled by the quotient of two differentials. Thus, unlike our objective point of view of the derivative, in this conception the derivative was not viewed as an immediate reflection of some external reality.

Modern analysis too has retained the objective point of view of the differential, though here it has acquired a different meaning. Here , first of all, we separate from among the totality of available variables, those which we consider to be independent, and the differential of such a variable dx is simply considered to be its arbitrary finite increment Δx. For the remaining variables, which are functions of the separated independent variables, some other definitions of the differential are well known; these definitions are different in form. For the sake of simplicity, we shall limit ourselves to a single function of a single variable x:

$$y = f(x).$$

Then according to Stolz's definition, used in the better textbooks of modern analysis (those of de la Vallée Poussin and Courant, for example) if $\Delta y = A \cdot \Delta x + \alpha \cdot \Delta x$, wherein A is not dependent on Δx, and α together with Δx tends to zero, then $dy = A \cdot \Delta x$.

Briefly, the differential dy is the principal linear part of the finite increment Δy. Other definitions of the differential are also accepted: according to the most widespread among them,

$$dy = f(x) : \Delta x;$$

according to Cauchy's definition,

$$dy = \frac{\lim f(x + h \cdot \Delta x) - f(x)}{h} \qquad h \to 0.$$

But when one intends to explain the most general meaning of these definitions, then one has recourse to Stolz's definition: it is not difficult to establish their equivalence with Stolz's definition.

There are deepgoing differences among the conceptions of the differential as an infinitesimal increment (in the meaning that was bestowed upon the words "infinitesimal increment" by the inventors of differential calculus), as an arbitrary finite increment or as its principal linear part; but all the same in all these cases, we deal with the objective point of view about the concept of differential.

In both the cases the differential immediately reflects some external reality, every time, just like the variables x and y themselves.

According to the second point of view, the *derivative*

$$f'(x) = \lim \frac{\Delta y}{\Delta x} \qquad \Delta x \to 0$$

immediately reflects some external reality; for the sake of brevity we shall call it the *operational* point of view.

Here the concept of differential reflects the well known aspects of those mathematical operations, from which the definition of the derivative and the computations with the derivatives follow.

From this point of view, the differentials are *introduced* in the form of ratios of differentials, ratios—that are *symbolized* in the *derivatives*:

$$f'(x) = \frac{dy}{dx}.$$

After this, it is not difficult to understand, that the operations with the symbolic ratios $\frac{dy}{dx}$, according to the very rules that are applicable to the algebraic fractions, will not lead to any contradiction. But neither is it mandatory, that we seek an *immediate* interpretation of each and every result that follows from these operations. In particular, nothing obstructs us from viewing the formula

$$dy = f'(x)\, dx$$

(obtained by freeing the aforementioned formula of the derivative of the denominator), as only another expression of the formula

$$\frac{dy}{dx} = f'(x).$$

Substantiation of the operational point of view had to wait for a considerably longer period of time, than what was required for, the substantiation of the objective point of view about the differential. The difficulty here was not with establishing the very *possibility* of thus, and not otherwise, interpreting the differential, but rather with the discovery of the *meaning* of such interpretation.

In this direction, the first methodologically exhaustive work was done by K. Marx; this work was written about fifty years ago, but was published only last year [K. Marks, *Matematicheskie Rukopisi* (Mathematical Manuscripts) // *Pod Znamenem Marksizma*, 1933, 1, str. 15-73]. J. Hadamard has treated the differential along the same lines, but differently, in his modern textbook [J. Hadamard, *Cours d'Analyse*, Paris, Hermann, 1927, pp. 2-10.].

2. The differential of Marx and of Hadamard.

The well known theorem about the differentiation of the function of a function will play the principal role in what follows: let, as before,

$$y = f(x);$$

and let x be the function of some other variable t; then

(1)
$$\frac{dy}{dt} = f'(x) \cdot \frac{dx}{dt}.$$

Marx's idea is as follows. When the starting point of the differential calculus, the equality

$$\frac{dy}{dx} = f'(x)$$

is taken isolatedly, then, at first, it has only a descriptive character. The real result of the operations, through which the derivative is determined, stands on the right hand side of this equality; the left hand side serves only as a symbol of those very operations, which lead to this result. But after the meaning of this symbol has been defined by such an equality, in consonance with this definition, it appears on both the sides of formula (1). [Hadamard proceeds from a formula, which is analogous to formula (1), but involves two variables: if

$$z = f(x, y),$$

then,

$$\frac{dz}{dt} = \frac{df}{dx} \cdot \frac{dx}{dt} + \frac{df}{dy} \cdot \frac{dy}{dt};$$

Marx has used a particular instance of this last formula: if $z = x \cdot y$, then,

$$\frac{dz}{dt} = y \cdot \frac{dx}{dt} + x \cdot \frac{dy}{dt}.$$

However, we lose nothing by illustrating their deduction by the simpler formula (1).] But there we are already dealing with operations under similar symbols: they themselves become the object of a calculus. "Thereby the differential calculus appears as a specific type of calculus, already independently operating on its own ground". Naturally, the formulae of this calculus have the meaning of *operational* formulae: thus, formula (1), upon establishing the connection between $\frac{dx}{dt}$ and $\frac{dy}{dt}$, thereby shows, what one must do, so that having $\frac{dx}{dt}$ one may obtain $\frac{dy}{dt}$. And from this very point of view this formula appears to be even more excessively complex. Since the nature of the variable t has no significance here, reference to this variable—to the differential dt—may be dropped altogether, just as one drops the common factor. According to Marx, this is not only possible, but also necessary: for the removal of the illusion, as though the formula (1) is true only in respect of some really independent variable. Thus we get a new formula

(2) $dy = f'(x)\, dx$.

This is an operational *symbol*: once we obtain the formula (2) as a result of computations, it is enough to divide both the sides of this formula by dx, for obtaining the derivative; it is enough to divide both the sides by dt—for obtaining formula (1) etc. Here the differential calculus as such finds its own natural fulfilment, and the "operational equation" (1), "as a preparatory equation, becomes superfluous, after it fulfils its task of supplying the general symbolic formula for differentiation" (2), "which directly leads us to our goal". [We may recall, that for Marx, strictly speaking, the equations (1) and (2) are not at issue; he was concerned with the equations

$$\frac{d(xy)}{dt} = y \cdot \frac{dx}{dt} + x \cdot \frac{dy}{dt}$$

and

$$d(xy) = y \cdot dx + x \cdot dy$$

But clearly, that does not change the affairs.]

Hadamard approaches the differential from the same operational point of view, but differently. Having established formula (1), he proposes to write out formula (2), by simply indicating this convention, and this alone, that whatever be the functional dependence of x and y on the parameter t, the equality (1) does hold good.

Hadamard insists on this definition of the differential, since it permits the use of the formulae containing operational symbols, for obtaining the derivatives. Let us assume, that computations with the differentials—in the sense that follows from Hadamard's definition—did indeed produce the formula,

$$dy = A \cdot dx.$$

Then one can assert that

$$f'(x) = A.$$

Actually, as per our assumption, for any t, we have:

$$\frac{dy}{dt} = A \cdot \frac{dx}{dt};$$

on the other hand, we know, that for any t:

$$\frac{dy}{dt} = f'(x) \cdot \frac{dx}{dt}.$$

From a comparison of these equalities, we get what is to be proved.

Hadamard defined the *second* differentials analogously. Proceeding from the formula

(1), Hadamard states the dependence between the *second* derivatives to be

$$(3) \qquad \frac{d^2y}{dt^2} = f'(x) \cdot \frac{d^2x}{dt^2} + f''(x) \cdot \left(\frac{dx}{dt}\right)^2$$

and proposes to write the formula

$$(4) \qquad d^2y = f'(x) \cdot d^2x + f''(x) \cdot dx^2,$$

as indicating this and this alone, that the equality (3) holds good, whatever be the functional dependence of the variables x and y on the parameter t. Formula (4) may again be used as the operational symbol for obtaining derivatives.

Let us assume, that computations with the differentials did indeed give us the formula

$$d^2y = A \cdot d^2x + B \cdot dx^2;$$

then we can assert that

$$f'(x) = A, f''(x) = B.$$

Indeed, as per our assumptions, we have for any t:

$$\frac{d^2y}{dt^2} = f'(x) \cdot \frac{d^2x}{dt^2} + f''(x) \cdot \left(\frac{dx}{dt}\right)^2.$$

From a comparison of these equalities, taking into consideration the fact that the derivatives of t can have any value whatsoever, we get what is required to be proved.

But is it necessary to abandon the definition of the differential as the principal linear part of an increment, in order to be able to thus use the formulae containing the differentials as operational symbols? This is not a very simple question and it requires to be discussed separately.

3. The operational point of view and the differential as the principal linear part of an increment.

We have seen, that the definition of the differential as the principal linear part of an increment demands, first of all, that we apportion one of the variables as independent. In reality, however, there are no absolutely independent variables. Even in the process of solving one and the same problem of geometry, mechanics etc., it is often impossible to consider one and the same variable independent, from the beginning to the end. It is clear, that the formulae of differential calculus, when applied to such problems, will really become full-fledged operational formulae, only if it is not required of us, that having once made a choice of an independent variable we should retain it in that capacity for the entire course of the computations; in other words, if the formulae, containing the differentials, and written with the assumption that x is the independent variable, remain unchanged, even when x turns out to be a function of another independent variable. In this sense, the invariance of the formulae, containing the differentials, thus happens to be an *essential condition* for the concordance of the objective and the operational points of view about the definition of the differential.

The definition of the *first* differential, as the principal linear part of an increment, satisfies this essential condition. Let us recall the proof of this well known fact. At issue here is formula (2): $dy = f(x) \cdot dx$.

When x is an independent variable, this formula is obtained as under. First of all

$$\Delta y = A \cdot \Delta x + \alpha \cdot \Delta x,$$

$$\frac{\Delta y}{\Delta x} = A + \alpha,$$

$$\lim \frac{\Delta y}{\Delta x} = A \qquad \Delta x \to 0,$$

$$f'(x) = A,$$

$$dy = A \cdot \Delta x,$$

$$dy = f'(x) \cdot \Delta x. \tag{2*}$$

Let us note, that in this argument the value of x is fixed, and Δx arbitrarily tends to zero: consequently, Δx is viewed as a variable, *not dependent on x*. This remark will play a vital role in what follows. As of now, let us return to what we have obtained:

$$dy = f'(x)\ \Delta x;$$

this apart, since $dx = \Delta x$, formula (2) quickly follows from it.

Now, suppose that x is not an independent variable, and $x = \Phi(t)$, such that it is also the case that $y = \psi(t)$. Then one may get convinced about the validity of formula (2) as under: Owing to (2*):

$$dy = \psi'(t) \cdot \Delta t;$$

according to formula (1):

$$\psi' t \cdot \Delta t = f'(x) \cdot \Phi'(t) \cdot \Delta t$$

and, finally, owing to (2*):

$$f'(x) \cdot \Phi' \ \Delta t = f'(x) \cdot dx \text{ [from (2*)} \cdot \text{we have } \Phi'(t) \cdot \Delta t = dx].$$

By comparing the last three equalities, we get formula (2). Clearly, (2) may serve as an operational formula, quite equivalent to Hadamard's definition.

Difficulties arise, when we try to define the *second* differential. The definition of the second differential is already included in the definition of the differential as the principal linear part of an increment, and there is no room for any additional "arbitrary" understanding. In fact, according to this definition, the differential dy is itself a function of x, and that is why the second differential d^2y, the differential of the differential, is thereby defined as $d(dy)$; it only remains to be computed. For this, let us note, that from the formula (2*) it follows that

$$d^2y = d\ [f'(x) \cdot \Delta x].$$

Further, as we have already noted, if we do not wish to make all our proofs of invariance of formula (2)—groundless, then we must view Δx as a variable independent of x. Hence, it easily follows, that the factor Δx stands after the differential sign, while differentiating from x, and we get:

$$d^2y = [df'(x)] \cdot \Delta x,$$

whence, having (2*) in view:

$$d^2y = f'(x) \cdot \Delta x^2. \tag{5*}$$

We get the final formula for the second differential, by substituting $dx = \Delta x$:

$$d^2y = f''(x) \cdot dx^2. \tag{5}$$

This formula, now found in the assumption, that x is an independent variable, turns out to be *non-invariant*. In fact, let,

$$x = \Phi(f), \quad y = \psi(t),$$

then, owing to (5*)

$$d^2y = \psi''(t) \cdot \Delta t^2;$$

according to formula (3)

$$\psi''(t) \cdot \Delta t^2 = f'(x) \cdot \Phi''(t) \cdot \Delta t^2 + f''(x) \cdot [\Phi'(t)]^2 \cdot \Delta t^2$$

and, finally owing to (5*)

$$f'(x) \cdot \Phi''(t) \cdot \Delta t^2 = f'(x) \cdot d^2x,$$

and owing to (2*)

$$f''(x) \cdot [\Phi'(t)]^2 \cdot \Delta t^2 = f''(x) \cdot dx^2.$$

Comparing the last four equalities we get:

$$d^2y = f'(x) \cdot d^2x + f''(x) \cdot dx^2.$$

This result does not coincide with formula (5).

The conclusion is clear. If we really want the differential calculus to be a full-fledged *calculus*, if we wish to have the right to use its formulae, as we use the algebra, without examining at every step, how they were obtained, then we shall be satisfied with the operational definition of the differential, as its basic definition. The concept of the differential as the principal linear part of an increment, turns out to be only an *interpretation*, suitable only for definite particular instances. When, in pursuit of an immediate and objective interpretation of each and every symbol, one accepts the principal linear part of an increment as the *definition* of the differential, i.e. when attempt is made to reduce the concept of differential as a whole to it, then one gets a defective result, since, by so doing, one fails to arrive at the differential calculus *as such*.

4. The operational point of view and the objective understanding of the differential in general.

Our conclusion may be explained only by stating, that, *namely, it is the operational understanding of the differential calculus, which reflects the reality correctly and fully* [even if, owing to the fact that in reality, as has already been mentioned, there are no absolutely independent variables].

Does this exclude any objective understanding, whatsoever, of the differential symbols?

Let us put the question more exactly: given a system of differential symbols dx, dy, d^2x, d^2y etc., mutually related—just as the derivatives of any variable t,

$$\frac{dx}{dt}, \frac{dy}{dt}, \frac{d^2x}{dt^2}, \frac{d^2y}{dt^2} \text{ etc. are—}$$

is any interpretation of these symbols, independent of the variable t possible?

We saw, that the interpretation as the principal linear part of an increment is possible only for the symbols of the first order dx, dy; from this, however, it does not follow that other interpretations, which would be suitable for the symbols of any order, are not possible.

Such interpretations do really exist. For example, *geometrical* interpretations are possible.

Let us take a curve defined by the equation:

$$y = f(x).$$

Let us assume, that on this curve

$$x = x(s),$$

$$y = y(s),$$

where s is the length of a segment of the curve (from a definite point and, it is positive or negative—depending upon the direction chosen). Now the differential symbols dx, dy, d^2x, d^2y may be interpreted as follows. Let us introduce some arbitrary constants not equal to zero, indicated through ds and d^2s and put:

$$dx = x'(s) \cdot ds,$$
$$dy = y'(s) \cdot ds,$$
$$d^2x = x'(s) \cdot d^2s + x''(s) \cdot ds^2,$$
$$d^2y = y'(s) \cdot d^2s + y''(s) \cdot ds^2.$$

It is easily seen, that this interpretation satisfies the equalities (2) and (4). The length s of the curve-segment, is a property of the curve—not dependent upon its analytical presentation. That is why the said interpretation too is not dependent on it. Naturally, whoever uses the differentials in this or that part of mathematics, bases his notion of the differential upon those interpretations, to which he is accustomed. Thus, interpretations like the one indicated are constantly followed in investigations pertaining to differential geometry, however, these are not created for the differentials themselves of the co-ordinates x and y, but rather for certain expressions formed with them, expressions—that are of geometric interest.

It is clear however, that the presence of this sort of interpretation does not, in essence, solve the problem of the differential. Marx solved this problem through a dialectical investigation of how the transition from algebra to the differential calculus was accomplished in mathematics. As a result of his investigations there arose the understanding of the differential calculus, as an algebra of its own kind, constructed over the ordinary algebra—which includes the differential symbols, besides the numbers. The definition of the differential provided in Hadamard's textbook shows that mathematicians are also arriving at that understanding of the general character of the differential calculus, where dialectics had arrived in the hands of a materialist philosopher—some half a century ago.

Source: Pod Znamenem Marksizma, 1934, 5. str. 79-85.

About the Author: Valery Ivanovich Glivenko (1897-1940), mathematician and logician; graduated from Moscow University in 1925, obtained his doctoral degree in 1928; taught in Karl Libknekht Teacher-Training Institute Moscow, from, 1928 to 1940; made great contributions to the intuitionist and constructivist logics.

Other Publications:

1. Sur la logique de M. Brouwer // *Bull. Acad. Sci. de Belgique* (5), 14(1928).
2. Logika Protivorechii. 1929.
3. Osnovy Obshiei Teorii Struktur. 1937.

For a complete list of his publications go to:
<http://www.mathnet.ru/links/62df1102ce4c2c6d74c11504439cd1a9/rm8831.pdf>

Marx's "Mathematical Manuscripts" and the development of history of mathematics in the USSR

Vladimir Nikolaevich Molodshii

The "Mathematical Manuscripts" of Karl Marx were published in 1968 [1], in connection with the 150-th anniversary of his birth. "All those manuscripts of Marx, which were more or less complete, or those which contained his own comments on this or that mathematical question", were included in this publication—in the original language and in Russian translation [1, s. 3]. A part of Marx's mathematical manuscripts, containing the results of his reflections on the nature of differential calculus, were published in 1933, in Russian translation [2].

For the Soviet historians of mathematics, the mathematical manuscripts of K. Marx, were important supplements to the fundamental works of the classics of Marxism-Leninism, upon which they constantly based their investigations. Here I have in view: Marx's "Capital", Engels' "Anti- Dühring" and "Dialectics of Nature" and, Lenin's "Materialism and Empirio-Criticism" and "Philosophical Notebooks". Marx's "Mathematical Manuscripts" helped the Soviet scholars to better orient themselves on philosophico-methodological questions—questions, that are important for the history of mathematics, which to some extent determined the concrete themes of their investigations, especially on the history of mathematical analysis and of its substantiation in the 17th-19th centuries.

Marx undertook a deeper study of mathematics in connection with his economic investigations [1, s. 4-6], His mathematical manuscripts show, that subsequently he became interested in purely mathematical problems—in questions pertaining to the problem of substantiation of the differential calculus, and in its history. Marx noticed the deficiencies of the basic conceptions of differential calculus of the end of the 17th-beginning of the 19th centuries and, he began to elaborate his own conception of "algebraic differentiation" and of the philosophico-methodological and historical questions intimately connected with it [1, s. 6-22]. Marx's conception is essentially different from that of Lagrange.

Marx treated the differential of a function as an operational symbol. In this connection he investigated questions related to the nature of mathematical abstractions and to its symbols, pertaining to making the definitions of the variable and the function more exact and, questions related to the mathematical means of describing movement. Marx discussed the question of regularity of the developments of mathematical conceptions in a "historical essay", in the light of the course of development of the differential calculus and the results of the attempts to substantiate it in the 17th-18th centuries.

The question of the nature of mathematical abstractions plays no small role even in the elaboration of the problems of the foundations of modern mathematics and of mathematical logic. It is enough to recall the struggle between the supporters and opponents of the concept of actual infinity. The dialectico-materialist elaboration of the problem of formation and of types of abstractions constantly drew the attention of some philosophers and mathematicians of our country—like S.A. Yanovskaya [3]. Facts from the history of calculi, axiomatic method and questions of mathematical logic were analysed in their investigations in a new light [4]. Comparisons with Marx's "Capital" were also undertaken. Marx himself had, on more than one occasion, indicated that the logic of mathematics and the logic of "Capital" resemble each other [5]; and this was not a chance remark. Yanovskaya used the results obtained in this field, while elaborating the question of the epistemological foundations of the criterion of truth in mathematics (the practice of non-contradictority) [6] and, those of the concept of mathematical rigor [7]. The creative output of S.A. Yanovskaya include the papers entitled: "On the so-called Definitions through Abstractions" (1935) and "The Problem of Introducing and Excluding the Asbtractions of Orders Higher than One" (1965) [7].

Modern investigators—especially philosophers and logicians—are drawn towards the question of operational strength of mathematical symbols. V.I. Glivenko was the first to offer an extended and purely mathematical evaluation of Marx's treatment of the signs dx and dy as operational symbols [8.1], (somewhat later, M. Fréchet too expressed analogous ideas) [8.2], Marx's ideas about the operational strength of these symbols are true even in respect of the wider range of materials provided by modern mathematics. As soon as the discussion turns to the scientific interpretations of the reasons behind the emergence and development of effective mathematical conceptions and methods—especially those of the 19th and 20th centuries, the aforementioned fact turns out to be important for history of mathematics.

Before the publication of Marx's mathematical manuscripts, historians of mathematics studied the ideas of the classics of Marxism-Leninism, on the inner-regularities of the development of the sciences, from the other sources available. But in Marx's "historical essay" on the development of the foundations of differential calculus, they came across Marx's own analysis of the inner regularities of the development of mathematics.

Marx showed that when the necessary conditions are created within the existing mathematical theories, then a new mathematical theory may arise and develop. This new theory "stands on its own legs", when its basic concepts and methods assume the specificities characteristic of it alone; the embryonic forms of these concepts contained in the initial mathematical theories, do not have these specificities. Marx stressed, that a new theory is not perfected and does not get recognition at once; that happens only through the struggle between its adherents and the followers of the old ideas.

Having compared three conceptions—those of Newton, d'Alembert and Lagrange—Marx observed, that in the period under consideration, the elaboration of the means of substantiating the differential calculus proceeded along the lines of perfecting and making things more exact.

The first investigations inspired by these ideas include: S.A. Yanovskaya's "Misheil Roll kak kritik analiza beskonechno malaykh" ("Michel Rolle as a critic of infinitesimal analysis") [first published in 1947, reprinted in: 3, s. 76-106]; and

K.A. Rybnikov's "Ob algebraicheskikh korniyakh differensialnovo ischeslieniya" ("On the algebraic roots of the differential calculus") [9], and "O roli algorifmov v istorii obosnovaniya matematicheskovo analiza" ("On the role of algorithms in the history of substantiation of mathematical analysis") [10].

Subsequently, not only the historians of mathematics, but also mathematicians and philosophers, began to take an interest in the question of inner regularities of the development of mathematics. Elaboration of this question was found to be essential for analysing the nature and mechanism of the scientific revolutions in mathematics.

Marx's "historical essay" helped to reveal the philosophico-methodological foundations of the mistakes and insufficiently substantiated conclusions of some of the leading mathematicians of the centuries gone by. Yanovskaya discussed the foundations of the mistakes of Saccheri's proof of the parallel lines axiom in: [11]. Rybnikov revealed the inexactitude in the arguments of I. Bernoulli and Ya. Bernoulli, in their analysis of one of the questions of variational analysis [12]. E. Ya. Bakhmutskaya compared the critical remarks of K. Marx and T.P. Osipovsky on Lagrarnge's conception of algebraic differentiation [13].

A study of the text-books of mathematics, published mainly in the 17th-18th centuries showed, that the process of making them more perfect at times gave rise to ideas, that went out of the framework of the then prevalent scientific ideas, but subsequently they became components of new conceptions together with a new methodology [14].

A.P. Yuskhevich noted [15], and then S.A. Yanovskaya and I.I. Likholetov showed [16], that from 1804 to 1834, the teaching of differential calculus in Moscow University went through three successive stages, reproducing its "mystical", "rationalistic" and "purely algebraic" forms respectively. Ideas of Cauchy replaced them by the middle of the 19th century. Roughly the same can be said about the teaching of mathematical analysis in the other universities and institutions of higher learning of Russia, during the first half of the 19th century [17 and 18].

Investigations of the Soviet historians of mathematics confirmed the correctness of Marx's critique of the conceptions of differential calculus held by Newton-Leibniz, d'Alemert and Lagrange, as well as of his position, that nevertheless, the elaboration of the questions of substantiation of differential calculus, as per these conceptions, went along a line of ascent (and that is why it produced concrete results). [The Leibniz-Newton apparatus of the differential calculus, together with its "principle of getting rid of" the infinitesimals of higher orders, has been provided with a new scientific substantiation in the non-standard mathematical analysis (see: "Matematicheskaya Entsiklopedia", T. III, M., 1982, s. 1019-1020. However, this does not reduce the merit of Marx's critique, since the Leibniz-Newton apparatus has been based upon essentially different presuppositions in the theories of its inventors on the one hand, and in the non-standard analysis on the other]. Marx's "historical essay" served as a starting point for those sections of some of the textbooks of history of mathematics, which contained a discussion of the development of mathematical analysis [19]. The second edition of the Great Soviet Encyclopaedia [20] and the "Filosofskaya Entsiklopedia" [21] contain a description of the contents and of the methodological significance of the mathematical manuscripts of K. Marx. Articles were devoted to these manuscripts in many journals, in particular in the "Uspekhi Matematicheskikh Nauk" [22], "Pod Znamenem marksizma" [23], "Voprosy Filosofii" [24] and in "Matematika v Shkole" [25], "Istoria Otchestvennoi Matematiki" [26] contains a brief description of the role of Marx's "Mathematical Manuscripts" in the development of history of mathematics in the USSR, during the last fifty years.

The questions about the stimuli, regularity and forms of scientific revolutions in the mathematics of 19th-20th centuries have principled significance for a dialectical materialist elaboration of the history of mathematics of the same period. Fruitful investigations of these questions are inseparable from the analysis of the law-governed development of the mathematical conceptions and theories as a single whole together with their basic concepts, principles, methods of proof and norms of mathematical rigor. Namely thus did Marx pose and investigate the question of development of the means of substantiating the differential calculus from the period of Newton and Leibniz to that of Lagrange. When the problem of scientific revolutions will gain its proper place in the investigations of our historians of mathematics, then they will be convinced about the fact that Marx's ideas about the regularities of development of mathematics, about the nature of mathematical abstractions and operational symbols, and about the struggle between the new and the old, are capable of helping them more than they did earlier. The present author became convinced about this while investigating the scientific revolutions in the theory of numbers of the 18th-beginning of the 19th centuries, and in the mathematical analysis and geometry of the first half of the 19th century. Other authors have also recognised and correspondingly used this fact in their investigations [27].

The ideas expressed by V.I. Lenin in his "Materialism and Empirio-Criticism" and in his "Philosophical Notebooks" "are also important for those historians of mathematics, who are engaged in the analysis of (scientific revolutions in mathematics—especially, of the period beginning with the end of the 19th century.

The last—50th—volume of the second edition of the collected works of Marx and Engels has been published recently. This edition includes 39 main and 11 supplementary volumes. This second edition contains nearly 800 new works and 700 letters more than those in the first [see: Sokrovishnitsa revoltsionnoi mysli (k zavershenіu 2-ovo izdaniya sochinenii K. Marksa i F. Engelsa), *Pravda*, 1983, 28 January]. In the 11 supplementary volumes, the historians of mathematics will find some new statements of Marx and Engels related to mathematics, to its role in the elaboration of the questions of the social sciences. This is of great significance for the study of the history of mathematization of knowledge in the 19lh century.

The entirety of Marx's manuscripts pertaining to the philosophico-methodological questions of mathematics and of its history, has not yet been published. These will be included in the multi-volume academic edition of the works of Marx and Engels, being prepared by the Institute of Marxism-Leninism of the CC of CPSU together with the Institute of Marxism-Leninism of the CC of SUPG [28]. However, a preliminary description of a part of this heritage has been published [29]; this publication may be used with profit.

Marx read (during 1878-79) Du Bois-Reymond's "Leibnizian Ideas in Modern Natural Sciences" [30] and noted his statement to the effect that: "... Aristotle's and Locke's view, that the soul is a *tabula rasa*, is supported by the investigations of Riemann, Helmholtz and others about the axioms of geometry" [29, s. 87]. It would be interesting to reproduce the rest of Marx's conspectus of this speech of Du Bois-Reymond and to compare it With the statements of Lobachevsky, Bolyai, Riemann, and Helmholtz about the nature of the presuppositions in geometry. A comparison of these notes of Marx with V.I. Lenin's conspectus of A. Rey's "Modern Philosophy" may also be of use. The following statement of A. Rey, noted by Lenin, in fact develops the aforementioned statement of Du Bois-Reymond, noted by Marx: "By constantly moving further from the space accessible to sense perception and by moving up to the geometrical space, mathematics, however, does not move away from real space, i.e. from the *true relations among things. But rather, comes closer to them*" [31]. At issue here are the mathematical abstractions characteristic of the mathematics of the second half of 19th century and, how the naturalists approached the dialectico-materialist interpretation of their nature. This is an important issue for those who are investigating the history of mathematics of the 19th-20th centuries.

Marx also took notes form the works of Leibniz and Descartes: from what they wrote about motion, and from the Leibniz-Clarke correspondence [32], and from some posthumous publication of Descartes [33]. An analysis of these notes would certainly help the historians of mathematics to understand the "Mathematical Manuscripts" of Marx better.

One should not forget, that as of now, the scientific writings, especially those remaining in the shape of manuscripts of Marx and Engels have not been collected in full. One cannot exclude the possibility of discovery of such new materials as may be of interest to the historians of mathematics [see: 28 and 29].

Mathematicians and historians of mathematics of other countries have also taken note of Marx's mathematical manuscripts. Noteworthy in this connection is an essay of D.J. Struik [34], where he has compared Marx's conception of "algebraic differentiation" with the conceptions of Cauchy and his successors. Svetoslav Slavkov's monograph on Marx's mathematical manuscripts [35] was published in 1963. The German and Italian editions of [the first part of] Marx's "Mathematical Manuscripts" (1968), were published in 1974 and 1975 respectively. H.C. Kennedy read a paper on Marx's mathematical manuscripts in the 15th International Congress of the History of Science [36]. Soviet scholars should analyse these materials and ascertain the nature of the influence exerted by the works of Marx upon the development of philosophico-historico-mathematical investigations in the world.

[This is a re-written and updated version of the paper read by the author at the Second School of History of Mathematics, Liepai, 3-10, VII, 1978. The first version of this paper was published in: *Istoriko-Matematicheskie Issledovania*, Vyp. XXVI. s. 9-17.]

Literature

1. *Marks K.*, Matematicheskie Rukopisi, M., Nauka, 1968 (pred., prim, i kommentarii S.A. Yanovskoi). 639 str. + risunki.
2. *Marks K.*, Matematicheskie Rukopisi // *Pod Znamenem Marksizma*, 1933, 1, s. 15-73; v sb.: Marksizm i Estestvoznanie, M. Politizdat. 1933, s. 5-61.
3. *Yanovskaya S.A.*, Metodologicheskie Problemy Nauki. M., Mysl, 1972.
4. Sbornik Statei po Filosofii Matematiki. M., Uchpedgiz, 1936,
5. *Dunaeva V.*, K voprosu o matematicheskom metode v "Kapitale" K. Marksa // *Voprosy Ekonomiki*, 1967.8, s. 18-30.
6. *Yanovskaya S.A.*, Soderzhateilnaya Istinnost i Formailno-Logicheskaya Dokazuemost v Matematike // Praktika i Poznanie, M., Nauka, 1973, s. 247-272.
7. See in [3]: O roli matematicheskoi strogosti v istorii tvorcheskovo razvitiya matematiki i spetsiailno o "Geometrii" Dekarta // *Istoriko-matematicheskie issledovaniya* (dalee: *IMI*), vyp. XVII, 1966. s. 151-184.

8.1. *Glivenko V.I.,* Poniatie Differntsiala u Marksa i Adamara // *Pod Znamenem Marksizma*, 1934, 5. s. 79-85.

8.2. *Fréchet M.*, Sur la notion de différentielle // J. Mathémat., t. 16., 1937, p. 233-250.

9. *IMI*, vyp. XI. 1958. s. 583-592.
10. Tr. In-ta istorii estestvozn. i tekh., 1957, 17, s 287-299.
11. *Yanovskaya S. A.*, O mirovozzrenii N. I. Lobachevskovo // *IMI*, vyp. IIII, 1950, s. 60-64; see: [3]. s. 107-149.
12. *Rybinkov K. A.*, Pervye etapi razvitiya variatsionnovo ischisleniya // *IMI*, vyp, II, 1949, s. 404-429.
13. *Bakhmutskaya E. Ya.*, Timofei Feodorovich Osipovsky i evo "Kurs Mathematiki" // *IMI*. vyp. V, 1952, s. 28-74.
14. See, for example: *V.N. Molodshii*, Osnovy yeheniya o chisle v XVIII-nachale XIX v. M.. Prosveshenie, 1963; *A.P. Yushkevich*, da Kuinya Zh. A. i problemy obosnovaniya matematicheskovo analiza // *IMI*, vyp, XVIII, 1973, s: 157-175.
15. *Yuskhevich A.P.*, Matematika v MGU za pervye 100 let sushestovaniya // *IMI*. vyp. I. 1948. s. 43-140.
16. *Likholetov I.I. i Yanoskaya S. A.*, Iz istorii prepodavaniya matematiki v Moskovskom Universitete (1804-1860gg.) // *IMI*. vyp. VII., 1955. s. 129-130.
17. *Lobachevsky N.I.*, Nauchno-pedagogicheskoe nasledie. M., Nauka, 1976.
18. *Galchenkova R.I.*, Matematika v Leningradskom (Peterburgskom) Universitete v XIX v. // *IMI*, vyp. XIV, 1961, s. 355-392.
19. *Rybnikov K.A.*, Istoriya Matematiki. 2-e izd. Izd. MGU. 1974.
20. BSE, izd. 2-e. T. 26, 1954, s. 496-498.
21. Filosofskaya Entsiklopedia, T. 3, 1964, s, 342-343.

22. *Rybnikov K.A.*, Matematicheskie Rukopisi K. Marksa // *Uspekhi matem. nauk*, 10:1 (63). 1955, s. 197-199. *Rozov N. Kh.*, Matematicheskie Rukopsi K. Marksa // *Uspekhi matem. nauk*, T. 23. vyp. 5. (143), 1968, s. 205-210.
23. *Yanovskaya S.A.*, O matematichesikh rukopisyakh Marksa // *Pod Znamenem Marksizma*, 1933, l, s. 74-115.
24. *Ruzavin G.I.*, Matematicheskie rukopisi K. Marksa i nekotorye problemy metodologii matematiki // *Voprosy Filosofii*, 1968, 12, s. 59-70.
25. *Kisileva N.A.*, Karl Marks i matematika // *Matematika v shkole*, 1969, 1, s. 5-10.
Molodshii V.N., O matematicheskikh rukopisyakh K. Marksa // *tam zhe*, s. 10-23.
26. Istoriya otchestvennoi matematiki, T. 4, kn. 2, Kiev. 1970, s. 460, 461, 465, 496; *A.P. Yushkevich*, O razviti isorii matematiki v SSSR // *IMI*, vyp. XI, 1958; *IMI*, vyp. XXIV, 1979.
27. *Rybnikov K.A.*, O tak nazyvaemykh tvorcheskikh i kriticheskikh periodakh v istorii matematicheskovo anliza // *IMI*, vyp. VII, 1954, s, 643-665.
Molodshii V.N., Osnovy ucheniya o chisle v XVIII i nachale XIX v. M., Prosveshenie, 1963;
Molodshii V.N., O. Koshi i revolutsiya v matematicheskom analize pervoi chctveorti XIX v. // *IMI*, vyp. XXIII, 1978, s. 32-55.
Molodshii V.N., O filosofsko-metodologicheskikh predposylkakh otkrytiya i razrabotki N.I. Lobachevskim neevklidovoi geometrii // Filosofskie nauki, 1980, 4. s. 75-85.
28. *Senekina O.K.*, F. Engels i istoriya nauki i tekhniki (po rukopisnym malerialalm F. Engelsa v Institute Marksizma-Leninizma pri Ts K K PSS) // *Voprosy istorii estestvoznaniya i tekhniki*. vyp. 3(32). 1970. s. 14-19.
29. *Krinitskii A.M.*, Rabota K. Marksa nad voprosami estestvoznaniya (soobshenie po neopublikovannym materialam) // *Voprosy Filosofii*, 1(3), 1948, s. 72-92.
30. *Du Bois-Reymomd E.*, Zwei Festreden in öffentlishen Sitzungen der kg. Preuss. Akademie der Wissenschaften, 1871.
31. *Lenin V.I.*, Poln. sobr. soch., T. 29, s. 482 [Eng. ed.. M.. 1961. V. 38, p. 418].
32. *Leibnitii*, Opera philosophica (ed. Erdmann), Berlin, 1840. pars prior, p. 774-775.
33. *Descartes R.*, Opuscula postuma phisica et mathematica. Amsterdami, 1701.
34. *Struik D.J.*, Marx and Mathematics // *Science and Society*, 1948, 12. No. 1. pp. 181-196.
35. *Slavkov S.*, K. Marks i niakoi problemy na matematikata. Sofia, 1963.
36. *Kennedy H. C.*, Karl Marx and the Foundations of Differential Calculus // *Historia Mathematica*, 1977. 4. pp. 303-318; Russian tr. with additions // *IMI*, vyp. XXVI, 1982.

Source: Voprosy Estestvoznaniya i Tekhniki, 1983, 2, pp. 29-34.
Author: Vladimir Nikolaevich Molodshii (1906-1986), Doctor of Physico-mathematical Sciences. Professor.

On the operational logical apparatus working in Karl Marx's *Capital* and *Mathematical Manuscripts*

Vasily Ivanovich Przhesmitsky

***I.** On the logic of Marx's "Capital", the logical apparatus of the concrete sciences and, the actual, real contradictions.*

The question of the operational logical apparatus working in Marx's "Capital" is topical, not only in connection with the necessity of defending Marxism and the foundations of its philosophy, but also in connection with the requirements of the concrete sciences, in particular, those of physics and mathematics. In these sciences, specific logical problems are arising continuously; these problems demand that we turn to a logic, which is more powerful than that ordinary, traditional logic, which has become "mathematical".

Following Marx and Engels we shall use the term "Logic" to indicate the science, wherein the "laws of human thought" are investigated or, which is "a discipline about the laws of the very process of thought " [Marx K., Engels F., *Collected Works*, Russian Edition T. 20, s. 91; T. 21. s. 316] (Eng. ed., v. 25, p. 84; Marx-Engels, On Religion (in Bengali), Progress, M., 1981 p. 258—Tr.). This discipline enters into all the fields of knowledge in the same way, as the method of MOVEMENT of thought from the problems to be solved to their true solutions, as "the method for seeking new results, for the transition from the known to the unknown", as "the method of investigation and of thinking" [ibid, T. 20. s. 138; T. 21, s. 303] (Eng. ed., v. 25, p. 125; Marx-Engels, On Religion (in Bengali), 1981, p. 288—Tr.). While studying the works of Marx and Engels, one cannot fail to notice, that already in the 1840s, they were using a qualitatively new (dialectical) logic which they themselves developed. After the publication of Marx's "A Contribution to the Critique of Political Economy", in 1859—they openly declared that a dialectical logic has been created, and that it is successfully working in science. In this connection, they noted, that the scientific result obtained by them in logic is," in its significance, hardly inferior to the fundamental materialist ideas" [ibid, T. 13, s. 497] (Eng. ed., v. 16, p. 475—Tr.). This result was obtained on the basis of the principles of philosophical materialism: "to comprehend the specific logic of a specific subject-matter" [ibid, T. I, s. 325],"... an understanding of nature as she is, without any extraneous additions..." [ibid, T. 20. s. 513] (Eng. ed., v. 25, pp. 478-479—Tr.), "one must not introduce any arbitrary sub-divisions" into the subject-matter under investigation, and the logical aspect of the subject-matter must "find its unity in itself" [ibid, T. 40. s. 10] (Eng. ed., v. Up. 12—Tr.).

Marx and Engels began their work in the field of logic, while they were still young. Already in 1837, Marx wrote confidentially to his father: "... this work ... which had caused me to rack my brains endlessly, (since it was actually intended to be a new logic) ... [it is] my dearest child" [ibid, T. 40. s. 15.] (Eng. ed. v. 1, p. 18—Tr.).

This statement indicates the indisputable beginning of the marxist [?—Tr.] dialectical revolution in logic. The new logic may be derived, with the help of strict logical means, from its primary element, its primary foundation, its initial "cell". Such an element must fixate the total overcoming of that one-sidedness of scientific thought—by logic—which emanates from the Aristotelian principle of non-antinomicity of truth. A study of Marx's and Engels' works show, that their purely logical demand is as under: to do away with the metaphysical anti-dialectical prejudice, the globalization of the principle of non-antinomicity and TO USE "IN THE PROPER CASES BOTH" THE DISJUNCTION $(v \dot{\vee} \overline{p})$ AND THE CONJUNCTION $(p \& \overline{p})$, AS PERMISSIBLE LOGICAL FORMS OF TRUTH, understandably, basing oneself upon the possible "use of the opposites" [ibid, T. 20, s. 528] (Eng. ed., v. 25, p. 493—Tr.), where it is necessary (our stress—V.P.).

Such an attitude to the conjunction $(p \& \overline{p})$ as a possible form of truth, fixated/consolidated in the Principles of logic as a science, is strictly scientifically substantiated. This (substantiation) consists of the two following facts, related to the DIALECTICS OF NATURE: (I) a contradiction in a scientific theory—that, which in the final count takes the form of the conjunction $(p \& \overline{p})$, (has) by no means (its origin only in thought) in any and every case; depending upon the concrete conditions, the concrete contents of thought, the foundation and the root of a concrete contradiction lies: either really in thought, or—in the perception of reality and only then in thought, or—in the very essence of the subject-matter of thought, i.e. exclusively in the very nature of things; (II) the set of three basic types of contradictions are divided in scientific theories, into concretely defined sub-sets or classes of fully defined specificities, these are strictly differentiated in the writings of Marx and Engels.

According to Marx and Engels, the first two types of contradictions constitute the class of "trivial" and, "false" or "seeming" contradictions and, they are to be contrasted with the remaining class of "REAL or TRUE" contradictions.

Through their investigations in the "Capital" Marx and Engels made it possible for us to understand the importance of the distinction between the real contradictions characteristic of reality and the contradictions of the sophist type, of that between the REAL contradictions and the really TRIVIAL (ILLUSORY, SEEMING) contradictions, those that give rise to the PARADOXES. It must be pointed out however that ordinary logic lacks the means required for visualising and expressing these distinctions, in so far as these distinctions are not immediately given by the forms of expressions, but rather by the contents thereby reflected. Since the logical form of all types of contradictions happen to be the one and same antinomy—the conjunction $(p \& \overline{p})$, the determination of whether this or that real concrete contradiction is genuine and true or an illusion or a paradox is impossible with the help of ordinary logic. For this logic even the task of distinguishing between a REAL contradiction from a sophistic-contradiction, becomes a BACK-BREAKING task in a number of instances.

Now let us take a look at the *real* contradictions of scientific theories.

We may take the following contradiction, as an example of the really false or seeming contradiction, that of the illusory contradiction or paradox, from amidst the writings of Marx and Engels: "why does the capitalist, whose sole concern is the production of exchange value, continually strive to depress the exchange value of commodities? A riddle with which Quesnay, one of the founders of Political Economy, tormented his opponents, and to which they could give him no answer" [ibid, T. 23, s, 331; *Capital* (Eng. ed.), I, p. 303]. The following example of a real, *genuine*, i.e. true contradiction may also be considered from among the writings of Marx and Engels: "Motion itself is a contradiction—even simple mechanical change of position can only come about through a body being at one and the same moment of time both in one place and in another place, being in one and the same place and also not in it. And the continuous origination and simultaneous solution of this contradiction is precisely what motion is" [ibid, T. 20. s 123; Eng. ed., v. 25. p. 111]. We may consider, alternatively, yet another example: "...it is a contradiction to depict one body as constantly falling towards another, and as, at the same time, constantly flying away from it. The ellipse is a form of motion which, while allowing this contradiction to go on, at the same time reconciles it" [ibid, T. 23, s 114; *Capital* (Eng. ed.), I, p. 106].

Remarkably, all these types and classes of contradictions, which we meet within the scientific theories, have been—one and all—investigated in the writings of Marx and Engels. The task of recognising and revealing the specificities of these contradictions has been made an ordinary, routine and everyday affair, like that of solving the riddles or revealing the secrets of nature or those of the objects of thought, which have often been termed—after Hegel—as the problems of "solving" (logical) contradictions. Such solutions are carried out by seeking out the essential, natural, intermediate, logically-mediating links. Incidentally, often many such intermediate links are required to be sought, as, for instance, from the standpoint of elementary algebra many intermediate terms are required, to understand that $\frac{0}{0}$ may represent an actual magnitude [ibid, T. 23, s. 326; *Capital* (Eng. ed.), I, p. 290].

The works of Marx and Engels also help us to understand the fact that the difference of the actual real contradictions from the really seeming and false ones—the paradoxes—which are BASIC to logic (and not to any theory of development whatsoever), leads to the following.

1) The real and actual contradictions in theory are antinomical-truths, as truths they may be logically used in appropriate situations either as natural elements of true scientific thought or as logical TRANSITIONS encountered in logical movement of thought—which the developed sciences cannot do without, whereas just like the sophistic-contradictions, the paradox type contradictions cannot play that role.

2) The real and actual contradiction characteristic of an object of thought, is contained in the very ESSENCE of the object or it may even be THAT VERY OBJECT (as in particular, is the contradiction of MOTION), while the really paradox producing contradiction is merely a contradiction between the outer side (and correspondingly, the external appearance) of the object of thought and its ESSENCE (and correspondingly, its inner regularity).

3) From the standpoint of ordinary logic, the real paradox generating contradictions, just like any trivial contradiction—cease to EXIST upon their resolution and, permit non-antinomical representation of the phenomenon indicated by them. In contrast to this, no actual, real contradiction (like the contradiction of MOTION) can be depicted non-antinomically and in this sense they are INDESTRUCTIBLE.

The actual, real contradictions do not vanish even in those situations, when together with their carriers—as is characteristic of them, they are DEVELOPED—i.e., transformed, unfolded and, are supplemented with new phenomena, with new contradictions derived out of the old ones. Since the development of an object DOES NOT ELIMINATE these real contradictions but only creates a *modus vivendi*, i.e. a form in which they can move forward side by side, a form of their movement And such in general is the method—explained Marx—with the help of which real contradictions are resolved [see: *Capital* (Eng. ed.), I, p. 106].

All that has been said above oblige us, following Marx and Engels, to see and to take note of the fact, in all the disciplines (and especially in logic and in philosophy as a whole), that the antinomies and the symbolic conjunctions which express them, may depict, not only the false and trivial contradictions, absurdities and paradoxes, but also the real, actual contradictions. And in that case such a conjunction can be the TRUE form of a TRUE antinomical proposition, VITALLY IMPORTANT FOR THE SCIENCE of logical transition along the path of dialectical negation together with a subsequent negation of the negation and, Marx has graphically demonstrated the justifiability and result-yielding character of such conjunctions in mathematics [see: Marks K., *Matematicheskie Rukopisi*, M., 1968, s. 29-75]. It must be clearly stated here, that now, when the works of Marx, Engels and Lenin are to a significant extent available to all, and are within the. reach of every Marxist, we should not fail to notice this fact.

Here it may be indicated as a positive fact, that mathematicians in the USSR have already noted, that the formal-logical type of thinking and the very principle of non-antinomicity may, under certain circumstances lead even "to the appearance of delusions and mistakes "[see: Rybnikov K.A., *Vvedenie v metodologiyu matematiki* (Introduction to the methodology of mathematics), M., 1979, s. 50]. And that is why, this type of thinking "did not and does not occupy a central position ... in developed human consciousness. The acquisition of mathematical knowledge, and its composition, includes within itself a lot of elements that are not amenable to formal-logical analysis. Quite a few of the methods of mathematical "operation" happen to be "non-logical" [Rybnikov, op. cit., s., 56-57]. And in so far as even to-day, we are required to use a "means of logical deduction which arose in the past" [ibid, s. 83-84], including the principle of non-contradiction as the principle of non-antinomicity, "notwithstanding the evident logical discrepancies and lack of explanations of the operational part of analysis", when "the problem of construction of models and their corresponding definite logical requirements have come to occupy the fore-court "[*ibid*, s. 94]. Here the development of mathematics (as well as that of theoretical mechanics and physics) is in ever greater need of a recognition of that DIALECTICS OF NATURE in logic, about the non-naturalness of which N.N. Luzin, D.D. Morduhai-Boltovskoi, P.S. Novikov and others have spoken repeatedly.

II. The operational logical apparatus of the "Capital" and of the "Mathematical Manuscripts" of K. Marx.

The operational logical apparatus adduced in the courses of traditional logic, called forth to ensure the logical movement of scientific thought, does not guarantee the banning of false identifications, antinomies or disjunctions and, even of logical arbitrariness in scientific thought. The logic of the "Captial" of Marx not only guarantees the banning of logical arbitrariness in scientific thought, but also ensures its successful development : reproduction of the essence of the object under investigation in all its contradictoriness, consistency of thought, its truth, broadness of the logical frame and of the genralisations adduced, the necessary strictness and, together with it, versatility of the conclusions. It is that is why, namely, that the logic of "Capital" is at least as superior to the ordinary logic "as the railways are to the medieval means of transport", to borrow an expression from Engels [Marx K., and Engels F., *Collected Works*, Russian Edition, T. 13. s. 48; Eng. ed. v. 16, p. 476).

The following logical devices have been employed in the "Capital" for ensuring the reproduction of the essence of the object of investigation in the movement of scientific thought: contradictions connected with the solution of the problems under investigation have been revealed and formulated—these contradictions reflect the essence of the problems to be solved. Such a logical order is clearly visible in Marx's formulation of the question of essence of commodity, of the capitalist form of social production and, that of the distribution of capitalist profit and capital [ibid, T. 13, s, 47-48; Eng. ed., v. 16, pp. 475-476].

Further, the contradictions, which have been formulated as fragments of the reality under investigation, and which appear as riddles or as secrets of the object of thought, are solved purely logically, i.e., in thought. This does not change anything in their content, yet removes the veil of secrecy from their face, removes the veil of mystery. In so far as the realisation of such an operation, while resolving the real, and even seeming, contradictions, i.e., paradoxes, demands that not only the external aspect, the appearance of phenomena, but also their internal side, the regularity, and that is also the essence of the OBJECT OF THOUGHT be taken into account; here thought, while realising the logical operations, moves from the formulation of the questions to their true solutions, without tearing itself off, either from the essence of the object of thought or, from the essence of the scientific problems to be solved.

Truth is the most valuable property of thought and, that is why, it is first of all the duty of LOGIC to control the truth of scientific thought. Since, namely, "LOGIC = THE QUESTION OF TRUTH" [Lenin V.I., *Collected Works*, Eng . ed. v. 38, p. 175]. While performing this duty, the logic of Marx's "Capital" —as distinct from ordinary logic—proceeds from the fact that concrete truths happen to be both non-antinomical and antinomical. That is why, as already stated above, the new logic, the logic of "Capital" —as distinct from the Aristotelian and modern formal logic—proceeds from the demand that BESIDES "EITHER-OR", "BOTH THIS—AND THAT" is also TO BE EMPLOYED in the RIGHT PLACE. At a definite stage, this demand wards off that situation, when actually true propositions are declared to be false on the basis of the fact that the positions opposed to them, formulated earlier, turn out to be true—and this is not a sufficient ground for declaring the earlier ones false. Here we are, in the main, speaking of the ready-made propositions and concepts. This is what I would like to point out at first.

Marx has also pointed out: "Truth includes not only the result but also the path to it. The investigation of truth must itself be true; true investigation is developed truth, the dispersed elements of which are brought together in the result" [Marx K., and Engels F., *Collected Works*, Eng. ed., v. 1, p. 113.]. That is to say, we must take care both of the consistency and of the starting points of the movement of scientific thought. The logic of "Capital" takes this into account. From the very beginning it directs the subject of investigation along a path which rejects all mysticism and scholasticism and, ensures "THE JOURNEY TO THE THINGS, AS THEY ARE, I. E. UPTO TRUTH" [Marx K., and Engels F., *Collected Works*, Russ. ed., T. 1, s. 29].

In view of the fact that the logic of "Capital" and of the "Two manuscripts on the differential calculus" of K. Marx permits in the right place, a recourse to truth equally with structural disjunction as well as with conjunction—the actualisation of the formal-logical kind of *consistency of the movement of scientific thought* becomes impossible. That is why, in this logic, the movement of scientific thought is actualised as a movement from the statement of the question to be answered, to a true answer of it. Here the very process of movement of thought is actualised in the form of a consistent unity of two—in a certain sense contradictory to the each other—dialectical-logical processes, appropriately termed as: the ascent from the concrete to the abstract and, the ascent from the abstract to the concrete.

The ascent from the concrete to the abstract leads to an unification and organisation of the scientific data, from there one may proceed to a true solution of the problem under consideration. These data are embodied in the necessary premises of the ascent to the concrete strictly-scientific abstractions, which are called the initial abstractions for the ascent from the abstract to the concrete. Initial abstractions, either as a kind of INITIAL "CELL" of the object of thought or as a kind of true answer of the question under consideration, or as a kind of "LOGICAL BRIDGE" over the problems to be solved, turn out to be suitable for such an ascent. We shall discuss these two types of initial abstractions and ascents in detail in another paper.

The ascent from the abstract to the concrete is an intellectually realised theoretical transition according to the corresponding laws of reality (these laws are logical and special-scientific); it is a transition from the already prepared strictly-scientific initial abstractions corresponding to the concrete scientific content, to the concrete knowledge sought—to the true answer of the question under consideration.

In order to ensure that the unity of the initial abstractions of the ascent, with the laws of the reality connected with them, was natural and, that they be dependable theoretical Elements and be the logical path of ascent to the true answer of the question under consideration—logic must take into account not only the external aspect, the appearance, but also the essence of the object or phenomenon considered. That is why dialectical logic must seek or elaborate the necessary concrete-scientific abstractions or laws of reality, initial to the ascents, only in indissoluble unity with the corresponding concrete sciences—and not otherwise. That is why in the "Capital" —where the object of investigation pertains to economics and history—this logic functions in an unity with political economy and history of mankind. In the "Two manuscripts on the differential calculus" —where the object of investigation, the differential, is a mathematical object—this logic functions in an unity with mathematics. It is understandable that this unity of the dialectical logic with a corresponding concrete science must take place only there and to that extent, where and to which extent it happens to be necessary.

The dialectical logic of Marx and Engels attains a broadness of logical frame and scientific generalisation, since it aims at using the unity of opposites: of the individual (the particular and the singular) with the general, of the concrete—with the abstract, of the real—with the possible, and of the determinate—with the indeterminate [Aristotle underrated the latter, but the indeterminate should not be underestimated] (see: Lenin V.I., *Collected Works*, Eng. ed., vol. 38, pp. 359-360). That is why, while rising from the concrete to the abstract, Marx reduces the general contained in the things to their most generalised logical expression [Marx K., and Engels F., *Collected Works*, Russ. ed., T. 3., s 180]. On this road one is required to raise scientific thought above the level of ordinary logic, i.e. above the level of the logical connections of identity and difference, of affirmation and negation, above the level of the Aristotelian laws of excluded middle and non-contradiction and, correspondingly of the formulae:

$$\neg(p \,\&\, \overline{p}) \text{ and } (p \dot{\vee} \overline{p}),$$

from among which neither one is for ever-true, nor for ever-false (but rather, both are only sometimes-true and sometimes-false), and thus rise to the level of connecting the opposites: AFFIRMATION and NEGATION, to express which one uses $(p \dot{\vee} \overline{p})$ together with $(p \,\&\, \overline{p})$.

This required representation of this law of unity of the opposites AFFIRMATION and NEGATION is superficially expressed more fully and deeply within the framework of its abstract universality, in the language of modern symbolic logic, by the formula:

$$\forall p \{[\neg p \,\&\, \neg \bar{p}] \dot{\vee} [p \,\&\, \neg \bar{p}] \dot{\vee} [\neg \bar{p} \,\&\, p] \dot{\vee} [p \,\&\, \bar{p}] \dot{\vee} [(p \,\&\, \bar{p}) \,\&\, (\neg p \,\&\, \neg \bar{p})]\}$$

In logic we shall call this formula and the concrete law which it expresses, Engels' logical law and formula of disjunction-conjunction without forgetting the fact that this is only one of the possible PARTICULAR expressions of the ESSENCE and "NUCLEUS" of DIALECTICS, of the law of unity of opposites.

In view of the universality of this law and of the formula that expresses it, while solving a concrete problem—the object of thought, characterized by a certain predicate in p and the predicate itself, is chosen in an extremely generalised form. That is to say, always only an individual indeterminate example is chosen, from amongst the set of all the examples of a given genus, as the object of thought, which potentially carries any form inherent to their identity and difference. This holds good also for the predicates which figure in thought as per necessity. Thus scientific thought protects itself from being suppressed PREMATURELY within the limits of the Aristotelian frame-work of such conditions as, the "one and the same object", at "one and the same time", in "one and the same relation" etc.; this provides the possibility of retaining, in the movement of thought from the statement of the question to its true answer, all its possible potentiality, which can not be immediately realised and utilised in full.

The logical means that regulated, as an instrument and as a method, the strictness and the versatility of scientific thought in the "Capital" and in the other writings of Marx and Engels, has the following poles.

When thought moves from a question to its answer, those boundaries are not to be lost sight of, within the limits of which, Engels' law and formula

$$S = \{[p \,\&\, \neg \bar{p}] \dot{\vee} [\neg p \,\&\, \bar{p}]\}$$

is adequate for the object of thought. Within these boundaries they (this law and formula) demand extreme concreteness from scientific thought (and thereby this thought becomes extremely strict). But this formula may be used only within the limits of its actual range of applicability (and herein the dialectical logic of Marx and Engels retains within itself all that is really valuable and true in the ordinary logic). The movement of thought from the statement of a question to its true answer—according to the law that connects the opposites within the frame-work of a necessary abstract universality, which is regulated within the frame-work of Engels' law and formula of disjunction:

$$\{[p \,\&\, \neg \bar{p}] \dot{\vee} [\neg p \,\&\, \bar{p}] \dot{\vee} [\neg p \,\&\, \neg \bar{p}] \dot{\vee} [p \,\&\, \bar{p}] \dot{\vee} [(p \,\&\, \bar{p}) \,\&\, (\neg p \,\&\, \neg \bar{p})]\}$$

by way of excluding from it every time those terms, which lose their meaning as per the given conditions and upon solution of a given problem—alone can ensure the extreme strictness and fluidity of thought.

Source: Voprosy dialekticheskoi logiki: Printsipy i formy myshleniya (materialy postoyanno deistvuyuschevo simpoziuma po dialekticheskoi logike). AN SSSR. Institut Filosofii, M., 1985. s. 70-81.

Author: Vasily Ivanovich Przhesmitsky (1913-), employed in the Department of Theoretical Mechanics of the Moscow State University of Civil Engineering since 1964. His scientific interests included dialectical materialism and its logical method, the principles of modern mechanics including dynamics of super-high-speed motions with their multi-sided problems requiring the use of logical methods. He retired in 1981.

On the problem of situating Marx's mathematical manuscripts in the history of ideas

Pradip Baksi

I

Karl Marx completed his school education in 1835, with "a good knowledge of mathematics" [20, 644], which included arithmetic, algebra, geometry, trigonometry and infinitesimal calculus [25, 157 ff]. However, he did not study mathematics in any university department. In the universities of Bonn and Berlin, he attended lectures on law, Greek and Roman mythology, Homer, history of modern art, anthropology, logic, geography, Isaiah and Euripides [20, 657-658 and 703-704]. While attempting an elaboration of a philosophy of law of his own, as a 19 year old student of Berlin University, he expressed his dissatisfaction, on a methodological plane, with "the unscientific form of mathematical dogmatism" [20, 12]. In the same year he composed three poems in jest, and gave them the common title: *Mathematical Wisdom* [20, 545-546]. Two years later, in 1839, he drafted a *Plan of Hegel's Philosophy of Nature* [20, 510-514], three versions of which have come down to us; they contain references to mechanics. *His Notebooks on Epicurean Philosophy* [20, 403-509], dating back to the same year and, his doctoral dissertation, written during 1840 and 1841 and submitted to the University of Jena in April 1841: *On the Difference Between the Democritean and Epicurean Philosophy of Nature* [20, 25-105], contain evidences of his continuing philosophical interest in the fundamental physico-mathematical concepts. Thus, in spite of a lack of formal mathematical education at the university-level, mathematics was always present in Karl Marx's intellectual horizon, in some form or the other, even during his formative years. In the latter half of the 1840's Marx's interest in mathematics was rekindled by the requirements of his investigations in the field of political economy. But within a few years this interest began to draw sustenance from other sources too: for instance, in July 1850 we find him discussing the then emerging materialist conception of nature and human history in the light of the developments in mechanics and in the other sciences [19, 67-69]; in April 1851 his friend Roland Daniels was imploring him to take up the study of physics in connection with the projected preparation of an encyclopaedia of the sciences [8, 113]; and in September-October that year Marx did study a treatise on the history of mathematics and mechanics (Marx, Karl (2018), *Mathematical Manuscripts:* 98-102. Delhi: Aakar Books.) After that he went through the different branches of elementary mathematics all over again and, made a special study of ordinary algebra and differential calculus. These studies continued for the next thirty odd years and, came to an end only with his death.

In "... mathematics Marx found the most consistent and at once the most simple expression of dialectical movements" [19, 31-32]. He was drawn to mathematics owing to the "many points of contact between mathematics, philosophy and dialectical logic" [16, 587]. In his more or less complete mathematical manuscripts he investigated the dialectic, the being and the becoming, the nature and history, of the fundamental concepts of differential calculus. That is why only in the context of the history of interaction of mathematical and philosophical thought can we hope to take our first steps towards a proper assessment of Marx's contributions in this field.

II

On the strength of the evidence available till date, it may be asserted, that mathematical activities began on our planet in the Neolithic era. It developed further in the centres of first urbanization: in ancient Mesopotamia, Egypt, China and India. This early mathematics was—to use a modern term—constructive, primarily oriented towards the construction of mathematical objects. The early texts containing arguments around mathematical objects and techniques grew somewhat later: in ancient Greece. Though some other civilizations—namely, the ancient Indian civilizations—had a formidable theoretical culture of their own, the interaction of mathematical and philosophical thought has been found to have been most prominent in ancient Greece alone.

It is commonly held that mathematics became theoretical with the geometrical investigations of Thales (624-548 B.C.). Thales was exposed to the mathematical practices of the ancient Egyptians, who in their turn inherited the mathematics of ancient Chaldea (Mesopotamia). Anaximandrus (611-545 B.C.) of the Ionic school founded by Thales introduced the concept of indefinite magnitude into mathematics. Another Ionic thinker Heraclitus of Ephesus (530-470 B.C.) was one of the pioneers in the conscious employment of dialectical reason in philosophy, in connection with the problem of conceptualization of change. These theoretical concerns gradually culminated in the first noticeable disquiet with the indeterminate magnitudes: in the reactions of the school of Pythagoras (580-504 B.C.) to numerical analysis. And finally, in the paradoxes associated with the name of Zeno of Elea (c. 490-430 B.C.), the problem of definite description of the indeterminate assumed explosive dimensions. These paradoxes have a close parallel in the problematique of *Milinda Prashna* associated with the name of king Milinda or Menander I (165/155-130 B.C.) [24].

The discovery of irrational proportions in the school of Pythagoras and the paradoxes formulated by Zeno led ancient Greek mathematics to the door steps of a "crisis of foundations". To meet this challenge Eudoxus of Cnidus (c. 390-337 B.C.) developed the method of exhaustion, already introduced by Hippocrates of Chios (c. 470-410 B.C.). In the Eleatic school the concern with the indefinite went side by side with the development of the *reductio ad absurdum* argument. The problems thrown up by the contradictions involved in the attempts at a definite description of the indefinite, generated attempts to demonstrate something by expelling the obvious formal contradictions. And thus the grounds were created for raising the question of dialectics of mathematics more categorically. This the Socratics did.

In Plato's (428/427-348/347 B.C.) *Republic* we find a Socrates dissatisfied with the lack of conceptual clarity of the empirics and the Pythagoreans, proposing an investigation into the dialectic of the fundamental concepts of the existing mathematical disciplines [27, 510-511 and 524-533]. Plato attempted to put his dialectics and mathematics on a common foundation in his theory of idea-numbers. An analogous approach may also be found in Aristotle's (384-322 B.C.) remarks on an universal mathematics [3, 42]. However, it was Aristotle's systematization of logic in the image of the existing Greek mathematical theory, which influenced later mathematical developments most strongly.

Euclid (approx. 300 B.C.) benefited greatly from the critical movement of Plato's academy, revising the principles of geometry (Eudoxus was associated with this movement). After that the method of exhaustion was further developed by Archimedes of Syracuse (c. 287-212 B.C.). Study of the conic sections was introduced earlier by Menaechmus and Plato, this was further developed by Apollonius (c. 262-190 B.C.) and his followers. With these developments ancient Greek mathematics reached its zenith.

In the period that followed, at first the Hellenic centres of the mediterranean countries (like Alexandria) became the repository of the ancient Greek attainments in mathematics. Afterwards the centres of mathematical and philosophical investigations in the Greek tradition shifted to Western and Central Asia. Meanwhile Chinese and Indian mathematics were not standing still. For a brief review of the mathematical attainments of the people of ancient and medieval Asia see: [31], [32], [33]. After the 6th century A.D. the Greek, Chinese and Indian mathematics interacted with each other, in the centres of learning of Arabic and Persian speaking Asia. In consequence we witnessed those developments in arithmetic, algebra, geometry, trigonometry and mechanics, without which the subsequent emergence of the differential and the integral calculus would not have been possible. During the 10th-13th centuries medieval southern Europe became aware of these achievements through Hebrew and Latin translations of the available literature in Arabic. This period also witnessed a renewal of European interest in ancient Greek Philosophy and literature as a whole. At around the same time, however, the living contacts between the mathematical traditions of Asia and Europe started getting snapped. The crusades (11th-13th centuries) provide the background to these contacts and their subsequent demise.

These contacts were to be re-established some four or five centuries later, as one of the results of the colonial expansion of the European capitalist powers in the East. By this time Europe became the centre of scientific activities. The new achievements of European sciences of the modern age trickled down to the littoral towns of the colonies through the distortive filter of the colonial educational policy of the European powers. However, even this meagre ration of new knowledge ushered in a process of renewal of learning in some of the colonies. The efforts of Tafazzul Hussein Khan and Raja Rammohun Roy in the realm of mathematics[1], indicate the beginning of this renewal in our own country. Marx was in general aware of this process, and of its limitations; see, for example, his article entitled *The Future Results of the British Rule in India* (written in 1853) [22, 29-34]. We have lived through and are still living through the consequences of this rule. Science education in the erstwhile colonies still remains a "lagging-behind-model" of the same in the advanced capitalist countries. Add to this lag of the present, the near total absence of awareness about our past attainments in the sciences. Those who study the history of ancient and medieval Indian sciences—a large part of which is occupied by mathematics—are promptly ticked out as purveyors of "soft science" and as nationalist propagandists. Of course the emergence of the study of history of science is connected with the rise of patriotic consciousness in our society, and it is a significant phenomenon in the history of the sciences in India, for that reason alone. But its role does not get exhausted just there, it is merely the first lap of a long and interesting journey. The study of the history of ancient and medieval sciences has profound contemporary significance. We shall mention just one example.

D.H.H. Ingalls (1951), *C. Goekoop* (1967) and *V.A. Smirnov* (1974) have, among others, indicated the need for further investigations into the *logic of relations* present in *Gaṅgeśa's* (13th century) *Tattvacintāmaṇi* [29]. Such investigations are already being undertaken and will be undertaken in future, with greater competence; these will provide us with a more complete picture of the logico-mathematical activities conducted on our planet, and consequently, with greater insights for dealing with the outstanding problems in this field. Here a pertinent problem may be posed (sometime in 1974-75, my teacher Pabitrakumar Roy made me aware of it, in Santiniketan): in India we had a long tradition both in mathematics and logic; the great names of *Gaṅgeśa* and *Bhāskarācārya* (12th century; infinitesimal method) are associated with them; and yet, the subsequent developments in mathematics and logic leading to the emergence of mathematical logic took place in Europe, and not in India—why?

It is customary to attempt an externalist answer to this (or to any other similar) question, in terms of the relative stagnation or dynamism of the modes of production. Such attempts are valuable (for sociology of knowledge) in their own right. But how are these traits (stagnation and dynamism) of the historically existing modes of production (e.g. Asiatic Mode of Production, Capitalism etc.) *mediated* into the sphere of production and reproduction of knowledge, into that of mathematical and logical thought in particular? We still do not have a definitive answer. As of now we can only propose a strategy for future investigations.

In ancient and mediaeval Indian mathematics the deductive aspect (arguments etc.) had a subsidiary position, the constructive aspect was preponderant [on this see: *Uspensky V.A.*, What is a Proof? // Theme No. 6 in: Reflections on seven Themes of Philosophy of Mathematics, in part three of this collection]. This resulted in a considerable development of the algorithms of arithmetic, algebra and trigonometry. In contrast, the deductive moment was very strong in ancient Greek geometry. Classical Greek logical theory was abstracted mainly out of this geometry. On the other hand, ancient Indian logical theories were, in the main, abstracted out of the ancient Indian prescriptive grammatical tradition, which again was constructive (algorithmic), with its stress on the construction of "algebraic" sutras for the normative conduct of linguistic activities. Perhaps this grammatical tradition itself served as the model for the ancient Indian mathematical disciplines. Perhaps F.J. Staal did not overstate the point when he asserted that: "Historically speaking the grammatical method of Pāṇini has been as fundamental for the Indian thought, as is the geometrical method of Euclid for the western [European] thought" [30]. The modern European developments in classical mathematics and logic, culminating in the emergence of mathematical logic towards the end of the last century, re-created the ancient Greek unity of the deductive moments of mathematics and logic, on a newer plane. It is only in this century that the constructivist[2] trend has begun to assert itself in the world of mathematics (and in the cognate disciplines) as a whole. And perhaps it is not accidental, that the Indians are again making their presence felt—this time through the technological end of the spectrum, in the field of computer softwares. Falling in line with the dominant tradition in history of mathematics and logic, so far, we have been studying the ancient Indian texts of mathematics and logic, through the methodological filter of classical (western) mathematics and logic. We may now attempt a *constructivist reading* of them. Hopefully, this time there will be lesser paradigm-mismatch.

Now, before we get back to the developments of mathematical and philosophical thought in medieval and modern Europe, we must mention, further, that in the period spanning from the fall of ancient Greek civilization right upto the emergence of the new bourgeois civilization in 15th century Europe, mankind witnessed several attempts to construct encyclopaedias of the sciences. Well known among these are: the *Naturalis Historia* of Gaius Plinius Secundus (23-79), in 37 volumes; the *Rasā'il al-Ikhwān al-safā'* [*Treatises of the Brethren of Purity*] in 53 volumes (10th-11th centuries); *Imago Mundi* of Petrus de Alliaco (1410) and, *Margarita Philosophica* of Gregor Reisch (1503). Direct and / or indirect cumulative impact of such attempts to collect the totality of existing human knowledge contributed to the regeneration of an interest in dialectics in medieval Europe. Thus, the grounds were prepared for a closer interaction between dialectical reason and mathematical investigations on a more advanced level. However, theology was the principal arena for the development of dialectical reason in medieval Europe. And consequently we find a Nicholas of Cusa (1401-1464), mixing the ideas of a dialectical logic based on religious mysticism "with the emerging notions of mathematical analysis" [5, 81].

With the rise of capitalism in Europe came its individualist philosophy dominated by rationalist metaphysics. In this very period the operations with variable magnitudes and functions were highlighted in mathematics, through the investigations of Descartes (1596-1650), Leibniz (1646-1716), Newton (1642-1727), Euler (1707-1783), d'Alembert (1717-1783), Lagrange (1736-1813) and their contemporaries and followers. Among them first Descartes and then Leibniz toyed with the idea of an universal mathematics. Such developments further widened the scope for investigations into the dialectics of mathematics. But the practising mathematicians took a different course; in a sense a natural course—that of exhausting the limits of formal reason. Bolzano (1781-1848), Dedekind (1831-1916) and Cantor (1845-1918) arrived at the conclusion that to deal with the fundamental concepts of the mathematics of variable magnitudes, i.e., of the differential and the integral calculus—the derivative and the integral—in a proper manner, infinite sets must be very precisely investigated. Cauchy (1789-1857), Weierstrass (1815-1897) and others developed the theory of limit. Consequently, the edifice of classical analysis was erected upon set theoretic foundations. But almost simultaneously, the well known paradoxes of the set theory arrived on the scene. Russell (1872-1970), among others, arrived at analogous paradoxes through his studies in mathematical logic. Thus the formal developments in classical analysis and mathematical logic prepared the grounds for a second "crisis of foundations" of mathematics (the first "crisis of foundations" was associated with Pythagorean numerical analysis and Zeno's paradoxes). In response there arose three schools: logicism, formalism and intuitionism. The arguments that followed led to considerable modifications, even abandonment, of some of their respective positions, in the wake of Kurt Gödel's (1906-1978) famous results about the incompleteness and inconsistency of even the most elementary formal systems [13]. Gradually, the emergence of the intuitionist, constructivist and non-standard analyses, brought an end to the monopoly of classical analysis, in the second half of this century. [But in the countries with a backward current mathematical culture, like our own (reflected in the terribly poor state of the investigations into the foundations, history and philosophy of mathematics in our country), classical analysis is often the only analysis "available" in the class-rooms of mathematics, till date.]

During this entire period of formal developments, spanning the whole of the 19th and nearly half of the 20th century, mathematical epistemology faced the famous antinomy posed by Kant (1724-1804): "the observable world is finite but we can not find its limits in space and time; therefore, the world is not finite but infinite, and there exists only the search for the limit according to the regulative requirements of reason" [5, 84].

The dualism inherent in this position was, to begin with, philosophically overcome within the idealist tradition of classical German philosophy itself, namely, in the dialectics of Hegel (1770-1831), wherein "... for the first time the whole world, natural, historical, intellectual is represented as a process, i.e., as in constant motion, change, transformation, development; and the attempt is made to trace out the internal connection that makes a continuous whole of all this movement and development" [10, 34]. However, having remained chained to idealism, even such a grand attempt shrouded itself in obscurity and mysticism. This mysticism left its obvious mark in all of Hegel's writings on the topical problems of the sciences, mathematics included.

We have the testimony of Engels to the effect that Hegel left behind him "numerous mathematical manuscripts" [3]. But in view of their non-availability[4], as of now, we are constrained to fall back upon the relevant chapters of his *Science of Logic*, in our efforts to take a look at Hegel's excursions into mathematics [15, 129-170 and 198-344]. For Lenin's comments on the same see: [18, 116-119]. Hegel attempted an explanation of the mathematics of his time, especially of the differential calculus. But owing to his predetermined idealist point of departure, this attempted explanation remained artificial from the very outset. He merely dressed up his categories, like the "Quality", "Quantum", "Determination" etc., with mathematical trappings, especially with the frills of differential-calculus-terminology. At best it was an expression of some of the concepts of his dialectics in a mathematical language, in so far as it was possible; but it could not lay bare the objective dialectics of mathematics. Hegel himself conceded that his attempt is merely a philosophical explanation of the existing mathematical practice [15, 319]. <This view needs to be tested in the light the larger and currently growing body of texts indicated in the bibliography in Appendix I.>

In contrast to Hegal, Marx (1818-1883) studied the nature and history of the concepts and symbols of differential calculus and concluded that they are operational. Forty four years after Marx's death, in 1927, Jack Hadamard, an intuitionist mathematician, arrived at similar conclusions regarding the general nature of the fundamental concepts of differential calculus [see: *Glivenko V. I.*, The Concept of Differential according to Marx and Hadamard // part two of this collection, first article]. The intuitionists like L.E.J. Brouwer (1881-1966) and many early constructivists were unaware of Marx's mathematical investigations. Even to-day the overwhelming majority of the mathematicians, philosophers and historians of ideas are unaware of them. A quarter <now half> of a century has elapsed since the publication of the 1968 edition of Marx's mathematical manuscripts. Why then this lack of awareness about them, even in the late 20th <and early 21st> century atmosphere of information boom? Why the responses to these manuscripts are so few in number?

The reader of these lines may have noted the long gaps in, and the protracted nature of, the history of the efforts to unravel the dialectics of mathematics. Well, such is reality: "...dialectical thought—precisely because it presupposes investigation of the nature of concepts themselves—is only possible for man, and for him only at a comparatively high stage of development (Buddhists and Greeks) and it attains its full development much later still through modern philosophy..." [11, 223]. It may further be mentioned in this connection that the stage of theoretic thought associated with the emergence of materialist dialectics was proceeded, among other things, by the following encyclopaedic attempts to classify the existing branches of knowledge: that of the French Encyclopédistes (1751-1765), under the editorship of Diderot and d'Alembert; Saint Simon's (1760-1825) incomplete attempt, and Hegel's attempt to philosophically sum up the results of the natural sciences of the Newton-Linnaeus school. The end of the positivist attempt of Comte (1798-1857) and the beginning of the attempts of Marx, Engels, Daniels, Schorlemmer etc. are near contemporaneous.

To go back to the "silence" around Marx's mathematical manuscripts: intuitionist mathematics and Marx-studies could not (in spite of certain proximity of Marx's position in mathematics with those of the intuitionists) join hands and study the mathematical legacy left by Marx. Emergence of the constructivist movement from within the intuitionist trend has perhaps created more favourable grounds. In its current phase the constructivist movement has rejected Brouwer's "semi mystical theory of the continuum" [4, 308], but has retained his general theory about the operational nature of the standard mathematical quantifiers and connectives.

So much by way of a contextual retrospective of the problem under consideration. Now, on to the problem itself.

III

A. To-day we are in a position to extend Marx's study of the differential calculus to the study of the symbolic calculi in general. We may investigate the dialectics of the various alternatives in set theory, analysis, topology and, theory of categories. Through a two-way association of Marx-scholars and mathematicians engaged in the study of Marx's mathematical manuscripts, the mathematical heritage left by Marx may some day get integrated into the mainstreams of mathematics, just as inside of a hundred odd years after the publication of the first volume of Marx's *Capital*, his critique of political economy got integrated into, and began shaping, the mainstreams of economic thought. <One possibly promising route is emerging at the interfaces of category theory and dialectics; see: the bibliography in Appendix I.>

B. However, the relevance of Marx's mathematical manuscripts does not get exhausted there. In his fundamental articles on the nature and history of differential calculus, Marx has shown that the transition from ordinary algebra to differential calculus proper, involves an *inversion of method*, expressed through the genetic development of the characteristic concepts and symbols of this calculus (differential, derivative, dx, dy, $\frac{dy}{dx}$ etc.) culminating in their future operational role (indicating strategies of the steps to be taken). I.S. Narsky has pointed out that the methodological scope of this discovery of Marx goes beyond the framework of mathematics—it applies to the meaning of signs in general [26, 156; quoted in: 7, 94]. Thus the theoretical apparatus of Marx's mathematical manuscripts is of relevance to the study of sign systems, to Semiotics[5]. Through semiotic investigations, its methodological relevance extends to the study of all the disciplines as sign systems, as "languages", i.e., as systems of articulation. Expectedly, such studies will reveal the characteristic "internal forms" of the different disciplines and their interrelations, and thus facilitate the ongoing process of integration of the sciences and disciplines. In this vast world of semiotic investigations into the structures and functions of disciplines, extended from the computer technology to the art forms, the various trends of mathematics also have places of their own, along with the ordinary languages and other sign systems. <See: the bibliography on mathematics and semiotics in Appendix II.>

C. In contrast to Hegel, who attempted a philosophical explanation of the existing mathematics, Marx proposed a new way of doing mathematics bereft of metaphysics, idealism, mysticism, obfuscation and sleight of hand. In others words Marx attempted to *change* the existing practice of mathematics, its existing reality. This attempt to change the existing stale of affairs of the sciences of this world—for example, of the classical political economy in the *Capital* and related texts, of the classical natural sciences in the *Dialectics of Nature*, and of classical analysis in the *Mathematical Manuscripts*—is what distinguishes the materialist dialectics from that of Hegel, and for that matter from all previous dialectics (in this connection let us recall Marx's celebrated eleventh thesis on Feuerbach). In so far as Marx and his friends were not contented with mere philosophical interpretations of the world and its sciences, the significance of their theoretic activities became trans-philosophical. They attempted a radical reconstruction of the entire structure of human knowledge.

This attempt began towards the middle of the 19th century, but its fuller contours are gradually came to light only in the second half of the 20th century, and will become clearer, hopefully, by the first half of the 21st century, when the publication of the *Marx-Engels-Gesamtausgabe* is projected to be completed. The first edition of the first volume of Marx's *Capital* was published in 1867. The first edition of its fourth volume (edited by K. Kautsky) was issued in 1910; but that edition of the fourth volume (*Theories of Surplus Value*) had many radical defects; and ultimately, the final (third) part of this fourth volume of *Capital* in the edition presently in use, was brought out only in 1962. <The publication of Division II of the *MEGA* containing the *Capital* and related preparatory work was completed in 2012.> Engels' *Dialectics of Nature* was first issued in 1925; but a more complete version of Engels' studies on the dialectics of the classical natural sciences of his time could be brought out only in 1973 [12]. The first incomplete edition of Marx's *Mathematical Manuscripts* came out in 1968. Marx's *Ethnological Notebooks* were brought out in 1972 [23]. The manuscripts of the first materialist attempt at investigating the dialectics of the interface of the biological and the social sciences, *Microcosmos: A Draft Outline of Physiological Anthropology*, prepared in 1850 by Marx's friend Roland Daniels (1819-1855), was brought out only in 1987 [9]. A large part of Carl Schorlemmer's (1834-1892) manuscripts on the history of chemistry remains unpublished. Nearly half of the notes and manuscripts of Marx and Engels[6] remain unpublished too; these include: a part of Engels' notes and manuscripts on the history of England. Ireland and Germany, history of philosophy and history of military science and, Marx's notes and manuscripts on agriculture, agro-chemistry, history of technology, geology, biology, physiology and, a part (nearly 400 pages) of his mathematical manuscripts. Publication of these notes and manuscripts and their translation into the various languages of the world will go a long way in developing our conception of the theoretical legacy of Marx. As of now, we have just begun to understand that at the level of cognitive activities, Marx' effort constitutes an encyclopaedic endeavour at changing the (then and even now) prevalent ways of cognising the world. One of the open questions about the Marx' attempts in the realm of human knowledge, is about their relation with the dominant scientific programmes[7] of our time, namely, with the atomistic, Cartesian, Newtonian and Leibnizian programmes.

It appears now, that the task of situating Marx's mathematical manuscripts in the history of ideas unfolds itself at many levels: at the level of completing text editing and publication, that of textual studies, that of situating these manuscripts within the structure and history of Marx' total output, that of situating them in the history of mathematical and theoretic thought, and that of situating the emerging conception of Marx' legacy in the history of scientific programmes. All these tasks are interrelated. And finally, we should not go about them in the spirit of an archivist of ideas or with that of a traditional historian of the past, but must take them up in the spirit of Marx, i.e. by simultaneously engaging ourselves in a comprehensive and systematic, case by case, synchronic and diachronic, study of all the newly emerged and emerging sciences and technologies of our time. In view of the gigantic strides of human knowledge since the days of Marx, and especially in late 20th and early 21st centuries these tasks have to be tackled by scientific collectives, aided by the latest attainments of the information processing technologies,

NOTES

1. Tafazzul Hussein Khan, the vakeel or ambassador of Nawab Asaf-ud-Daulah at Calcutta during the Government of Marquis Cornwalis (1738-1805) translated into Arabic Apollonius' *De rationis sectione*, Newton's *Philosophia Naturalis Principia Mathematica*. Thomas Simpson's *Algebra* and William Emerson's *Mechanics*, during 1788-1792 [2, 39-40].

   Raja Rammohun Roy (1772-1833) wrote a modern treatise on *geometry* in Bengali [6, 407]. These manuscripts remain untraced till date.
2. On *Constructivism* see: note 111a to Marx's *Mathematical Manuscripts* in the present volume and, *Nepeyvoda N.N.*, Emergence and Development of the Concept of Constructivisability in Mathematics // Present Volume, part three, last article.
3. F. Engels wrote to F.A. Lange on March 29, 1865: "I cannot leave unnoticed a remark you make about old Hegel, who you say lacked the more profound kind of mathematical and natural-scientific training. Hegel knew so much mathematics that none of his pupils was equal to the task of editing the *numerous mathematical manuscripts* he left behind. The only man I know who understands enough mathematics and philosophy to do this is Marx" (my emphasis—P.B.) [21, 173].
4. In response to a personal inquiry from the present author, Dr Helmut Schneider of the Hegel Archiv. Ruhr-Universität Bochum, had informed in his letter dated February 7, 1982, that neither does the Hegel Archives possess the originals of Hegel's mathematical manuscripts, nor were they ever published. <On some of the later developments in this area see, Present Volume, Appendix I: A bibliography on Hegel and mathematics.>
5. *Semiotics* is the study of signs (Greek *semeion* = sign). A sign is a sensorily perceptible material object, action or, event, which denotes or represents another

object. Semiotics has its origins in some of the ideas of the American philosopher and logician Charles Sanders Peirce (1839-1914) and the Swiss linguist Ferdinand de Saussure (1857-1913). The range of semiotic investigations are extended over all patterned communication systems: from the simplest signal systems, through the ordinary languages used by people, right upto the special languages of various disciplines. These are traditionally divided into three parts: syntactics, the study of structure; semantics, the study of meaning; and pragmatics, the study of actual use. The growth of these investigations have given rise to prominent American French, Italian, Czech, Polish, Estonian and Russian schools of Semiotics. But in spite of both intensive and extensive developments the problem of Constructing a synthetic conception of sign remains open. Such a conception must answer the question of genesis of signs, as well as that of the generic functions of signs within a system. So far the *genetic* and *generic* approaches to the sign systems have been posed as "either-or" alternatives. A synthesis of these approaches within a "both-and" perspective is very much on the agenda of semiotic research. The mathematical manuscripts of Marx may provide this relevant perspective, in view of the fact that they contain such a conception of the characteristic signs of one language, that of the differential calculus. In this context also see: Piatigorsky, A.M. and D.B. Zilberman (1976), "The emergence of semiotics in India: Some Approaches to Understanding *Lakṣaṇā* in Hindu and Buddhist Philosophical Usages", *Semiotica*, 17, 3 (1976): 255-265; and, Appendix II: A bibliography on mathematics and semiotics.

6. For a brief outline of the contents of these notes and manuscripts see: [1], [17], [25] and, [28]. For the details visit: <http://mega.bbaw.de/struktur/abteilung_iv>
7. The concept of *scientific programme* has evolved out of the modern investigations in history of science, philosophy of science and science of science. The basic tenets of a scientific theory, its major premises, are always formulated within the framework of a scientific programme, which sets the ideal of scientific explanation and organisation of knowledge, and also formulates the conditions under which knowledge is considered to be both authentic and proved. The origins of this concept can be traced back to Imre Lakatos (1922-1974) ("research programme") and Thomas Kuhn (1922-1996) ("paradigm"). It has been further elaborated in subsequent research and alternative concepts of scientific programme have been proposed. A prominent worker in this field Piama Pavlovna Gaidenko (1934-) wrote on this concept:

   "So what is a scientific programme, and why has the need arisen for such a concept? Unlike a scientific theory, a scientific programme, as a rule, lays claim to cover all phenomena and to provide an exhaustive explanation of all facts, i.e., to an universal interpretation of everything existing. A principle or a system of principles formulated in a programme, is, hence, *universal in nature*. The well-known tenet of the Pythagoreans—"all is number", is a typical example of the concise formulation of a scientific programme. A scientific programme is most frequently formulated within the framework of philosophy (it is no accident that

Engels speaks of the principled impossibility for the natural sciences "to free themselves from philosophy") [see: 11, 209-210]. Its creators are scientists who, at the same time, also come forth as philosophers: after all a philosophical system, unlike a scientific theory, is not inclined to distinguish a group of *its own* facts, but lays claim to the universal significance of its principle, (It is precisely in analysing the structure of scientific programmes and the forms of their ties with the scientific theories, as well as in examining the evolution and change of programmes, that philosophy and history of philosophy can and must help history of science in solving its tasks).

Nevertheless, a scientific programme is not identical with a philosophical system or a definite philosophical trend. Not every philosophical system can produce a scientific programme. A scientific programme should contain not only the characteristics of the object under examination but also the, closely connected with these characteristics, possibility of elaborating a corresponding method of research. Thus, a scientific programme, so to say. determines a definite method of building a scientific theory by providing the means for the transition from a general world-outlook principle advanced in a philosophical system, to the revelation of the lies between the phenomena of the empiric world. Thus, three diverse scientific programmes came into being on the basis of ancient [Greek] philosophy: atomistic (which found its realisation in scientific theories only in modern times); mathematical (Pythagorean-Platonic, which already found its realisation in ancient times—in Euclid's *Elements* and in the mechanics of Archimedes); and finally. Aristotle's continualist programme, on the basis of which the first physical theory—the physics of the Peripatetic school—came into being. The major scientific programmes in modern times were created by Descartes, Newton and Leibniz.

*** *** ***

A study of the emergence evolution and, finally, death of scientific programmes, the emergence and consolidation of new programmes, as well as changes in the types of ties between programmes and the scientific theories based on them makes it possible to reveal the internal ties between science and that cultural-historical entity within the framework of which it exists. Such an approach also makes it possible to trace the historically changing nature of these lies, i.e., to show how the *history of science* is internally linked with the *history of culture*.

The fact that within a definite historical period not one but two and even more scientific programmes can exist side by side, which, according to their initial principles, are opposed in each other, does not allow for a simplified conclusion about the contents of these programmes, relying on some "primary intuition" of the given culture, but *calls for a more thorough analysis of the "composition" of that culture, of the diverse tendencies coexisting in it.*

On the other hand, the existence of more than one programme in each period in the development of science shows that the idea that the history of science is an uninterrupted, so to say "linear", development of definite originally set principles and problems is unjustified. The very problems which are being tackled by science, change in the course of its history: each historical period sees their essentially different interpretation" [14, 134-137].

Such in brief is the meaning and significance of the concept of scientific programme.

REFERENCES

1. *Baksi P.*, Karl Markser Prakriti Bijnan Charcha O Bijnan Bhabona [Karl Marx's studies of and thoughts on the natural sciences] // *Mulyayan*, Sharadiya, 1395 (October, 1988), pp. 157-174 and, Nababarsha, 1396 (May, 1989), pp. 74-80.
2. *Bandopadhyay S.*, Bangiya Ranesance Paschatyavidyar Bhumika [The Role of Occidental Learning in Bengali Renaissance). Calcutta. 1980.
3. *Beth E.W., Piaget J.*, Mathematical Epistemology and Psychology, Tr. by W. Mays. Dordrecht. 1966.
4. *Bishop E.*, The Constructivisation of Abstract Mathematical Analysis // Proceedings of International Congress of Mathematicians (1966). Introduction. Moscow. 1968 (pp. 307-313).
5. *Bogomolov A.S. et al*, Dialectical Logic // Themes in Soviet Philosophy. Selection of articles from *Filosofskaya Entsiklopediya*; Tr. by T.J. Blakely. Dordrecht. 1975.
6. *Chattopadhyay N.*, Mahatma Raja Rommohun Roy. Allahabad. 1928.
7. *Chepikov M.G.*, The Integration of Science. Moscow. 1978.
8. Daniels-Marx Correspondence; published under the title *Novye materialy o K. Markse* [New materials on K. Marx] // *Voprosy Filosofii*, No. 5. 1983, pp. 100-126. For a Bengali translation of these letters see: Marksbad o Bijnan Samuher Dvandvikata [Marxism and the dialectics of the sciences], Calcutta. 1986, pp. 17-84.
9. *Daniels R.*, Mikrokosmos: Entwurf einer physiologischen Anthropologie (1850) // Hrsg. von H. Elsner; Annot. und Beitr. von J. Bleker *et al.* Frankfurt a.M. etc., 1987.
10. *Engels F.*, Anti-Dühring. Moscow. 1978.
11. *Engels F.*, Dialectics of Nature. Moscow 1976.
12. Fridrikh Engles o Dialektike Estestvoznaniya [Frederik Engels on the dialectics of the natural sciences]. Ed by B.M. Kedrov. Moscow. 1973.
13. From Frege to Gödel: A source Book in Mathematical Logic. 1879-1931. Ed. by Jean van Heijenoort. Harvard University Press. 1967.
14. *Gaidenko P.P.*, Evolution of Science: The Cultural-Historical Aspect // *Social Sciences*, Vol. XIII, No. 2, 1981, pp. 131-144 For greater details, see by the same author:

    (1) Evolutsiya Poniyaliya Nauki: Stanovlenie i Razvitie Pervykh Nauchnykh

Programm [Evolution of the Concept of Science: Formation and Development of the First Scientific Programmes]. Moscow. 1980.

(2) Evolutsiya Poniyaliya Nauki (XVII-XVIII vv.): Formirovaniye Naychnykh Programm Novovo Vremini [Evolution of the Concept of Science (17th-18th centuries): Formation of the Scientific Programmes of the Modern Age]. Moscow. 1987.

15. *Hegel G.W.F.*, Science of Logic. Volume One. Tr. by W.H. Johnston and L.G. Struthers. London 1929.
16. *Karl Marx: A Biography*. Moscow. 1973.
17. *Krinitsky A.M.*, Rabota K. Marksa nad voprosami estestvoznaniya (soobschenie po neopublikovannym materialam) [K. Marx's writings on the questions of natural science (a communication based upon unpublished materials)] // *Voprosy Filosofii*, 1(3). 1948. pp. 72-92.
18. *Lenin V.I.*, Philosophical Notebooks. Collected Works. Eng. ed., Vol. 38. Moscow. 1961.
19. Marx-Engels Smriti [Marx-Engels: Reminiscences]. Moscow. 1976.
20. *Marx K., Engels F.*, Collected Works. Eng. ed. Vol. 1. Moscow. 1975.
21. *Marx K., Engels F.*, Selected Correspondence. Moscow. 1965.
22. *Marx K., Engels F.*, The First Indian War of Independence: 1857-1859. Moscow, 1978.
23. *Marx K.*, The Ethnological Notebooks (studies of Morgan, Phear, Maine, Lubbock). Ed. by L. Krader. Assen. 1972. (Published under the title: The Ethnological Notebooks of Karl Marx.)
24. *Milinda Prashna* [Questions of Milinda]. Bengali tr. of *Milinda pañho* by Pandit Shreemat Dharmadhar Mahasthavir. Calcutta. 1977.
25. *Monz H.*, Karl Marx: Grundlagen der Entwicklung zu Leben und Werk. Trier, 1973 [quoted in: *Reiprich K.*, O filosofsko-estestvennonauchnykh issledovaniyakh Karla Marksa (On the philosophico-natural-scientific investigations of Karl Marx) // *Voprosy Filosofii*. 1983. 12. p. 4].
26. *Narsky I.S.*, Dialectical Contradiction and the Logic of Cognition (in Russian). Moscow. 1969.
27. *Plato*, The Republic. Tr. by B. Jowett // The Dialogues of Plato (in two volumes). New York. 1937.
28. *Senekina O.K.*, F. Engels i istoriya nauyki i tekhniki (po rukopisnym materialam F. Engelsa v Institute Marksizma-Leninizma pri Ts K K P S S) [F. Engels and the history of science and technology (according to the manuscripts of F. Engels available in the Institute of Marxism-Leninism of the C C C P S U] // *Voprosy Istorii Estestvoznaniya i Tekhniki*. vyp. 3(32). M., 1970. pp. 14-19.
29. *Smirnov V.A.*, On the Reconstruction of the Navya-Nyaya // *Indo-Soviet Studies*, II. 1991. pp. 59-63 [Eng. tr. of Redaklorskie slova. *Vvedenide v Indiskuyu Logiku*. M., 1974).

30. *Staal F.J.*, Twee metodische richtÿnen voor de filosofie [quoted in: *Renou L.*, Panini // *Current Trends in Linguistics*, V. 5. 1969. p. 497: quoted in: *Paribok A.V.*, On the Methodological Foundations of Indian Linguistics // *Indo-Soviet Studies*, I, 1990, p. 19]. <For a different opinion on this issue see: Johannes Bronkhorst, The relationship between linguistics and other sciences in India, in: Sylvain Auroux *et al* Ed. *History of the Language Sciences...*, Volume 1 (2000): 166-173. Berlin/New York: Walter de Gruyter.>
31. *Volodarsky A.*, Soviet Studies on the History of Eastern Science // The History of Science: Soviet Research. Vol. II. Moscow. 1985.
32. *Yushkevich A.P.*, History of Mathematics of the Mediaeval Orient: A Survey of and Perspectives for Research // *Indo-Soviet Studies*. II, 1991, pp. 36-55.
33. *Yushkevich A.P.*, Mathematics and its History in Retrospective // Present volume, part three, first article.

The differential calculus, mathematicians and economists in the nineteenth century: K. Marx and H. Laurent, readers of J.L. Boucharlat

Alain Alcouffe

In 1881, two old men in London, German emigrants, without any particular mathematical training, were exchanging reflections on the differential calculus. That these men were Marx and Engels, and that these reflections were the result of an enormous amount of reading and elaboration by Marx over a period of nearly thirty years, would not be enough to make these exchanges escape from the caricature immortalized by Flaubert in *Bouvard et Pécuchet*[1] (and published in 1881, after Flaubert's death), as if these "mathematical manuscripts" were nothing more than fatuous[2]. These manuscripts contain about a thousand pages of mathematics written in instalments from 1846 or at least 1852 to his death in 1883. In this research, some have wanted to see preparatory work on economics, but it seems that Marx relentlessly took a real interest in mathematics itself, probably as an extension of his interest in Hegel's texts on differential calculus. Indeed, the *Mathematical Manuscripts* of Marx, apart from the interest they may have for the marxologist or philosopher of sciences, make interesting contributions, both direct and indirect, to the history of mathematics in the nineteenth century. Their interest is, in fact, twofold: on the one hand, Marx's reading of the history of differential calculus and, on the other hand, they are an incentive to read again his sources, which we can compare with other contemporary authors and place in the history of mathematics. From this point of view, one of Marx's sources, Jean-Louis Boucharlat's *Elements of Differential Calculus and Integral Calculus*, occupies a prominent place. It is, first of all, one of the most frequently used references by Marx, in particular, in his most elaborate text (the memoir on the differential calculus written for Engels, which they may have intended to publish[3]). There is no indication as to the reasons for Marx's interest in this particular book, a rather late choice apparently, since Marx

1. Cf. G. Flaubert, *Bouvard and Pecuchet: a Tragi-Comic Novel of Bourgeois Life*. The Floating Press. 2016, http://public.eblib.com/choice/publicfullrecord.aspx?p=4585572. The book details the adventures of two Parisian copy-clerks, that decide to move to the countryside. Their search for intellectual stimulation leads them, over the course of years, to flounder through almost every branch of knowledge Flaubert uses their ridiculous quest to expose the hidden weaknesses of the sciences.
2. The (most) complete edition of Marx's Mathematical Manuscripts was published in 1968, in a bilingual German-Russian presentation (Moscow : Nauka). It strives to present Marx's work in a systematic way, for the most part, readings and book reviews. The most elaborate passages were published in West Germany in 1974, (facsimile of the 1968 edition). Italian and English translations were published in 1975 and 1983 before the French translation of 1985. Two Chinese translations came out from Shanghai and Beijing in 1975. An Indian English translation and a Bengali translation came out from Calcutta in 1994.
3. Engels wrote to Marx on August 18, 1881 "You need have no fear that in all this you will have been anticipated by any mathematician.". Engels stressed also Marx's original discoveries in mathematics in his Speech at the grave of Marx, March 17, 1883.

had previously used a book by another French mathematician, L.B. Francoeur[4], but it must be admitted that he had made a choice ratified by many people concerned with this field of mathematics: here is a book first published in 1813 which was translated into many languages (including a translation into Hindustani), published in many editions without major revision, and still be reprinted more than a century after its first publication (until 1926 at least!). We propose to examine the presentation of concepts in Boucharlat's differential calculus , taking as a guide, on the one hand, Marx's criticism and, on the other hand, the 1886 version of Boucharlat's book which includes a revision of the treatment of differentials by Hermann Laurent.

I) General presentation of J.L. Boucharlat's *Elémens de Calcul différentiel et de Calcul Intégral*

Jean Louis Boucharlat was born in Lyon in 1775 to a wealthy family who gave him a solid education[5]. At a very young age, he developed a strong interest in both poetry and mathematics. Worried during the Terror, he took refuge in Saint-Etienne but then, equipped with letters of recommendation for Monge and Lagrange, he arrived in Paris, took the Polytechnique exams and was admitted. After graduating, he obtained a chair in mathematics in Lyon. He published mathematical works including a *Theory of Curves and Surfaces of the Second Order* "with the help of Lagrange and dedicated to him" (1810). In the meantime he had been appointed assistant repetitor[6] at the Ecole Polytechnique, from which he resigned, unhappy to have been ousted when the chair he taught was assigned to another.

Thanks to Lagrange's support, he finally obtained a chair in mathematics at the Prytané de la Flêche, whose mathematical course (artillery division) had been elevated to the level of that of Polytechnique. It was precisely the notes prepared for this course that were to constitute the *Elemens de Calcul Différentiel*, the first edition of which appeared in 1813. Under the Restoration, he resigned any function in the teaching of mathematics to devote himself to literature. He taught the literature course at the Athenaeum as an extension of La Harpe[7]. He died in Paris in 1848.

In his biography of 1846, *Theory of Curves* is presented as an introduction to the Elemens, while the *Elemens de Mécanique* published in 1815 would be a sequel. The revisions Boucharlat carried out in subsequent editions show that Boucharlat's interest in mathematics was not linked to his employment as a professor of mathematics but to a genuine intellectual inclination.

4. L.B. Francoeur, *Cours complet de mathématiques pures*, Paris, Bachelier, (3rd edition, 1828).
5. We follow the fairly consistent biographies of E. Garay de Monglavé (1846), the *Grande Encyclopédie* (1893) and the *Dictionnaire de Biographie Française* (1954).
6. A repetitor was an adjunct lecturer, usually the first step in an academic career at Polytechnique.
7. The Athénée of Paris was a 'learned society' founded in 1781 under the name of Lycée. It became the Republican Athénée, then the Royal Athénée. La Harpe, (1739–1803), a French playwright, writer and literary critic. taught there a *Cours de Littérature* which has been considered as " the emergence of the discourse of the history of ideas ". Cf. Goulemot Jean-Marie. Le Cours de littérature de La Harpe, ou l'émergence du discours de l'histoire des idées. In: *Littérature*, n°24, 1976. pp. 51-62

Hermann Laurent, born in 1841, was the son of the chemist Auguste Laurent[8]. He graduated from the Ecole Polytechnique in 1862. In 1865 he was awarded a doctorate in mathematical sciences with two theses entitled *De la continuité dans les séries* and *Sur les lignes isothermes* and was appointed, the following year, as a repetitor for analysis at the Ecole Polytechnique. After resuming his service during the War of 1870,[9] he left the army again in 1872, but he had to continue his life as a repetitor at the Ecole Polytechnique. From 1883 onwards, he performed the tasks of admission examiner and from 1889 onwards he was also professor of mathematics at the Institut Agronomique. He died in 1908. Hermann Laurent had an intellectual production of the first rank: he wrote many articles in the *Journal de Liouville*, the *Proceedings of the Academy of Sciences*, etc.. He also published several mathematical works among which we will examine in particular a *Traité d'Analyse* in 7 volumes (1885-91) which allows us to better situate the revisions he introduced in the 9th edition of Boucharlat. He was equally interested in statistics and financial mathematics and hence in political economy. Thus he was in contact with Léon Walras and published in 1902 a *Petit traité d'économie politique mathématique*, in accordance with the precepts of the Ecole de Lausanne.[10]

Although he was not credited with exceptional discoveries, he was nevertheless a very representative author of mathematics in the last quarter of the 19th century in France. This last point is not restrictive: H. Laurent in his *Traité d'Analyse* quotes Weierstrass and "verbatim translates" his demonstration that function, $f(x) = \cos \pi x + b \cos a \pi x + \ldots\ldots + b^n \cos a^n \pi +$ although continuous has no derivative. H. Laurent adds a note to his translation of Weierstrass stating that "these lines were written several years ago, at a time when many geometers refused to accept the previous conclusions which seem generally accepted today" (P.414). C. Méray, a French mathematician who shared with G. Cantor, Weierstrass and Dedekind the invention of the modern definition of numbers, was to congratulate H. Laurent on the earliness of his "rallying" to the new conceptions.[11]

II) The History of Differentials

Before examining the concepts of differential calculus in Boucharlat, it is interesting to carefully read his short preface in which he presents a periodization of the history of differential calculus, a periodization on which Marx had to rely. He distinguishes

8. Biographical information on H. Laurent is taken from the Grande Encyclopédie (H. Laurent was responsible for mathematical articles) and from the obituary published in 1908 in *L'Enseignement Mathématique*.
9. Polytechnique was created during the Revolution as a training school for the army. French graduates are automatically officers and serve for a period at least equal to the length of their studies in the army.
10. Laurent's book was published in 1902, Paris, C. Schmid. The relations between Laurent and Walras were essentially epistolary; see *Correspondence of Léon Walras and related Papers*, edited and annotated by W. Jaffé, N.H. 1965.
11. Cf. C. Méray, *Leçons nouvelles sur l' analyse infinitésimale et ses applications géométriques*, Paris, Gauthier-Villars, 1894. On the history of numbers at the end of the 19th century, see N. Bourbaki, *Eléments d' histoire des mathématiques*, Paris, Hermann, 2nd edition 1969. On the relationship between this history and the differential calculus, cf. C.H. Edwards, *The Historical Development of the Calculus*, Springer, New York, Berlin, 1979.

three periods. First of all, he locates the origin in Newton and Leibniz, inventors of an "analysis far superior to Descartes' geometry". But there were still remaining some "mysteries". Newton, the first one, penetrated this mystery by considering the differential calculus as the method of the first and final ratios of the quantities or otherwise as the method of the limits of their relationship. The method of limits was perfected by D'Alembert so that it was "freed from all consideration of motion, ideas foreign to the differential calculus. Finally, Lagrange's method in turn complements that of limits by attaching the differential coefficients to pure Algebra". In this sketch of the history of the differential calculus, the dominant impression is that of a very modern guideline, the passage from one stage to the next being linked to an "instillation of rigour", to quote Kline[12], but the end of the preface blurs this image, in fact, for Boucharlat, "these three methods form, so to speak, only one". This point puzzled Marx who rejected Boucharlat's eclecticism, seeking in the perspective of a Lakatosian internal history the reasons for the succession of the three methods.

It is not possible to compare the history of the differential calculus of Boucharlat-Marx with that of Laurent, because on the one hand, the latter purely and simply eliminated the preface of Boucharlat from his edition of the *Elements* (the 9°). On the other hand, Laurent has expressed his views on the origins of the differential calculus in passing comments in his correspondence with Walras and in articles in *L'enseignement mathematique*. It can be inferred from this that he shared the idea of a growing instillation of rigour: "When Leibniz created this admirable instrument called the differential calculus, he was unfairly criticized, and the proof is that differential notation is now universally adopted, but the criticisms that were directed at him had their raison d' être."[13]

But before establishing a periodization of the history of the differential calculus and the articulation of the different conceptions, it is necessary to examine one by one the concepts of the differential calculus to see the movement that can connect the different approaches. For this reason, it seemed preferable to give a direct reading of Boucharlat's presentation of differential calculus before confronting it with the readings of Marx and Laurent, on the one hand, and the evolution of differential calculus, on the other hand.

12. Mr. Kline entitled his chapter on the XVIII° "Instillation of rigour", *Mathematical Thought from Ancient to Modern Times*, New York, Oxford University Press, 1972.

13. H. Laurent," Considerations sur l' enseignement des mathématiques dans les classes de spéciales en France" in *L'enseignement mathématiques*, 1899; (Laurent opposes "the notation of Lagrange's accents, born out of a false idea, a barbaric notation rejected by all civilized peoples, rejected by Lagrange himself, who never used it in his memoirs, and the notation of Leibniz' differentials, universally used, as the simplest, most expressive and most convenient." p. 41); and in the correspondence Walras-Laurent (in Jaffé, letter 1452), Laurent writes to Walras: "My opinion is that in your theories, there is still a lot of vagueness, which may be said without offending you because one cannot demand more from a science such as yours, more than one has demanded from the founders of differential calculus. You have had the merit of posing problems and providing undoubtedly imperfect solutions from the point of view of rigour, as Leibnitz did". The quotation from the text is taken from the *Petit traité d'économie*, 1902, p. 3.

III) The concepts of function and limit

(Boucharlat, 1826, pp. 1, 2, 3).

DIFFERENTIAL CALCULUS

On the differentiation of algebraic quantities

1. One variable is said to be a function of another variable, when the first is equal to a certain analytical expression composed of the second: for example, y is a function of x in the following equations :

$$y=\sqrt{a^2-x^2},\ y=a^3+3bx^2,\ y=\frac{x^2}{a},\ y=b+cx^3$$

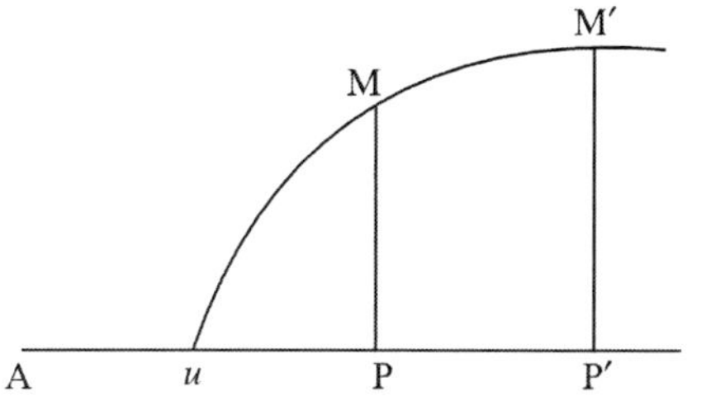

Fig. 1

2. Let us consider a function when in its state of increase, by reason of the increase of the variable which it contains ; and since every function of a variable x may be represented by the ordinate of a curve BMM′, Fig. 1, let AP = a and PM = y be the coordinates of a point M in that curve, and suppose that the abscissa AP receives an increment PP′ = h, then the ordinate PM will become P′M′ = y'. In order, therefore, to obtain the value of this new ordinate, we see that we must change x into $x + h$ in the equation of the curve, and the value which the equation shall determine for y will be that of y'.

For example, if we had the equation $y = mx^2$, we should obtain y' by changing x into $x + h$, and y into y', and we should have $y' = m(x+h)^2$.

Or, by developing $y' = mx^2 + 2mxh + mh^2$

3. Let us take also the equation $y = x^3$ ...(1)

And suppose that when x becomes $x + h$, y becomes y', we shall have then

$$y' = (x + h)^3;$$

Or by expanding, $y' = x^3 + 3x^2h + 3xh^2 + h^3$;

If from the equation we subtract (1) there will remain

$$y' - y = 3x^2h + 3xh^2 + h^3;$$

And by dividing by h,

$$\frac{y'-y}{h} = 3x^2h + 3xh^2 + h^3 \qquad (2)$$

Let us see now what is to be learnt from this result : y'-y, being the difference between the new value of y and its primitive one, represents the increment of the function y in consequence of the increment h given to x; and the increment of x, on the other hand, being h, it follows that the expression $\frac{y'-y}{h}$ is the ratio of the increment of the function y to that of the variable x. By attending to the second side of equation (2), we see that this ratio is diminished the more h is diminished, and that when h become 0 this ratio is reduced to $3x^2$. This term $3x^2$ is therefore the limit of the ratio $\frac{y'-y}{h}$, being the term to which it tends as we diminish h.

Boucharlat's formulations are very characteristic of early 19th-century mathematics. Thus, the definition of the function remains very similar to that given by Bernoulli in 1718: "Here, the function of a variable quantity is called a quantity composed in any way whatsoever of this variable quantity and constants", taken up by L. Euler in 1730, who was to develop it in 1748 into "an analytical expression made up of the variable and constants".[14] In particular, it does not distinguish between the function (relationship between two variables, one of which depends on the other) and the dependent variable. In $y = f(x)$ the "function" for Boucharlat is "y" and not "$f(\ )$".

Marx was interested in the evolution of the concept of function and the 1968 edition of the MM included the translation/transposition into German of the history of the concept he had found in Lacroix' Treatise (1810), as well as an analysis of the concept of function used by Lagrange. Lacroix did not mention any particular mathematician in his historical sketch, but presented the successive "developments" of the concept from the "ancient analysts (who) generally included under the term function of a quantity all the powers of that quantity" (Lacroix, 1810, p. 1) to the algebraic and transcendental functions. Lacroix was particularly interested in the development of a function in series which he contrasted with the value of a function, a distinction which testifies to the dissociation between "y" and "$f(\ )$"[15]. Marx tried to theorize this distinction by opposing 'function of x' and 'function in x': the former designating y's dependence on x, the latter a pure expression of x. He pointed out precisely the additional generalization in this area carried out by Lagrange and the subsequent confusion of meanings of the term. He notes that:

For Lagrange, on the other hand, the expression $f(x)$, as soon as it is found in the left side of the algebraic expression in x, refers only to the *general* expression, hence *indeterminate* as opposed to the *particular expression*; and $f(x + h)$ refers to the meaning of the general undeveloped expression as opposed to its developed expression of development in series, such as for example in algebra, $(x + a)^m$ is

14. J. Bernoulli, *Opera Omnia*, t. II, Lausanne and Geneva, 1742, p. 241, cited in A.P. Juschkewitsch, p. 241. "Euler und Lagrange über die Grundlagen der Analysis", in *Sammelband zu Ehren des 250. Geburtstages Leonhard Eulers*, Akademie-Verlag, Berlin, 1959.

15. S.F. Lacroix, *Traité du calcul différentiel et du calcul integrentiel*, seconde édtion, Paris, Courcier, 1810.

the general, undeveloped expression, whereas on the right side, we have the development in series, x^m + *etc.*

This is quite satisfactory and appropriate in certain well-defined cases, but the distinction between the function of x and the function in x cannot yet be emptied of its content because it alone implies that the function of x can have a concrete existence differentiated from the function in x, such as that of the ordinate if x is the abscissa [...]. (Marx, 1968, pp. 502-4)

But if the proposed distinction was intended to clarify the "confusions" between the multiple meanings of the word function, it does not seem to have found any application in the differential calculus where Marx uses an elliptical formula "function x".

For Laurent, it will be noted first of all that he repeats without modification in the 9th edition of Boucharlat the formulation given in earlier editions. On the other hand, in his *Traité*, he felt it necessary to start with some discussion of the definition of the term function. After having pointed out that the term has seen its meanings changed, he gives "the more precise definition" which it seems to him should be adopted and which he will adopt in what follows: "Two quantities are a function of each other when one remains constant, the other also remains constant." (Laurent, 1881, p. 1). This definition seems to be directly inspired by that of Cauchy, whose disciple Laurent always proclaimed himself to be; indeed, Cauchy presents the functions in the following way:

When variable quantities are so closely linked that, given the value of one variable, the values of all the others can be inferred, one usually conceives of the various quantities expressed by means of one of them, which then takes the name of an independent variable; and the other quantities, expressed by means of the independent variable, are what are called functions of that variable. (Cauchy, 1829, p. 5).

Laurent himself points out some of the disadvantages of his own definition, which erases the distinction between dependent and independent variables and further increases confusion. It is this situation that Frege referred to in 1891 in a paper entitled "Function and Concept" when he stated that the usual definition of function was not satisfactory because "no distinction is made between form and content, sign and thing signified [Bezeichnetes]; a mistake, admittedly, that is very often met with in mathematical works, even those of celebrated authors"[16].

This imprecision is reflected in the definition of the limit we find in paragraph 3 of the Boucharlat text. But here the delimitation of concepts has immediate consequences on the implementation of the differential calculus. It was Newton who had introduced the idea of limit using what Marx called a "metaphysical and non-mathematical hypothesis". Newton defended himself against those who objected to "the first and last sums and ratios of nascent and evanescent quantities" by

16. G. Frege, "Funktion und Begriff", reprinted in *Funktion, Begriff, Bedeutung*, Vandenhoeck & Ruprecht, Göttingen, 1969. [Peter Geach and Max Black eds. (1952). Translations from the philosophical writings of Gottlob Frege (1st ed.). Oxford: Basil Blackwell.]

specifying that it was necessary to understand "by the ultimate velocity is meant that with which the body is moved, neither before it arrives at its last place and the motion ceases, nor after, but at the very instant it arrives; that is, that velocity with which the body arrives at its last place, and with which the motion ceases".[17] The whole mystery seems to reside in this moment when the movement ceases. More precisely, if we give, as Boucharlat does, to an indeterminate variable x an increase h and we make $x + h$ tend to x, does "h becomes null" mean anything other than "the 'limit' is reached for $h = 0$" ? We know that Lagrange's project was precisely to establish the differential calculus without evanescent quantities, limits and the infinitely small i.e. by resolutely placing itself at point $h = 0$. On this point, it seems that Boucharlat's synthesis was purely verbal: he uses the term "limit" but in reality it is placed in the perspective of Lagrange and "h becomes null" simply means: $h = 0$.

This "ambiguity" of Boucharlat's notion of limits was the starting point of Marxian criticism. Marx notes, indeed, in the case of x^3 that:

If one argues that the more the value of h decreases on the right-hand side, so much the more does the value of the terms $3xh + h^2$ decrease, so that the value as well of the entire right-hand side $3x^2 + 3xh + h^2$ more and more closely approaches the value $3x^2$, we then must add, however : 'yet without being able to coincide with it'. [...] In this sense $3x^2$ becomes the limit value of the series $3x^2 + 3xh + h2$ [...]

However as soon as we set $h = 0$, the terms on the right-hand side vanish, making $3x^2$ the limit of its value; now $3x^2$ is the first derivative of x^3 and so = $f'(x)$. As $f'(x)$ it indicates that an $f''(x)$is also derivable from it [..] and thus that the increment $f'(x)$ or $3x^2$ is not =the sum of the increments which can be developed from $f(x) = x^3$ [...] In this sense, however, the developed series of increments becomes, as soon as I break it off, the *limit value* of the development [...] This has absolutely nothing in common with the limit value in the first sense (Marx, 1983, pp. 124-5)

It is this ambiguity in the post-Lagrangian use of the concept of limit that will lead Marx to reject it as misleading and constantly misleading. More deeply, we can see here a consequence of Marx's concern to deal with infinity because, on the one hand, he insists on the possibility that h decreases without limit, and, on the other hand, Marx wants to reason by posing $h = 0$ that is, to master the actual infinity. In this sense, one can assume that he would not have been satisfied with Cauchy's reasoning, even if he had known Cauchy, which does not seem to have been the case. He would undoubtedly have appreciated the way in which Cauchy clarified the concept of limits, but he avoided dealing with infinite quantities in the following way.

When the values successively assigned to the same variable approach indefinitely a fixed value so as to end up differing as little as one wishes, the latter is called the limit of all the others[18].

17. Cajori, F., & Newton, I. (1934). *Sir Isaac Newton's mathematical principles of natural philosophy and his system of the world.* Los Angeles, Calif, University of California Press, p. 39.
18. Augustin Cauchy, *Leçons sur le calcul différentiel*, Paris, Debure, 1829, pp. 1, 2.

At this point in examining the concepts of the differential calculus, it may be useful to verify that it is not just a matter of words, but that the implementation of different concepts leads to significant differences in demonstrations, some of which may be wrong. To this end, we choose Boucharlat's study of $\dfrac{\sin h}{h}$, a study that attracted Marx's attention.

39. *The arc is greater than the sine, and less than the tangent.*

To prove this, let AB, Fig. 2, be an arc, which has BE for its sine, and DA for its tangent, and take the arc AB′ equal to the arc AB. Then the chord BB′ being a straight line, BB′ is less than the arc BAB′; and therefore the straight line BE, which is the half of the chord BB′, is less than the arc BA, the half of arc BAB′; whence it follows that the sine is less than the arc.

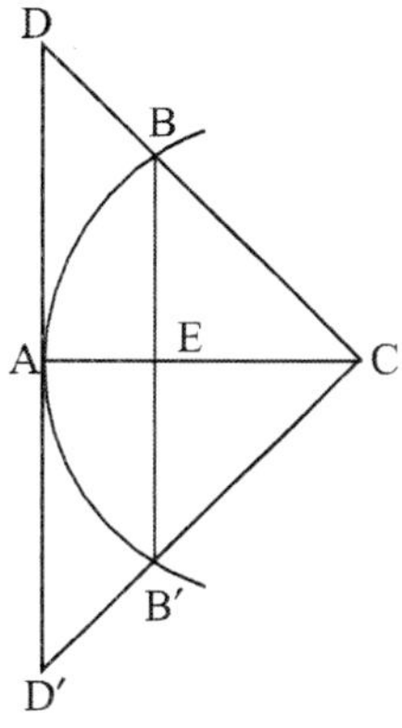

Fig. 2

To prove that the tangent is greater than the arc, we have

Area of triangle DD′C $>$ *area of sector BAB′C;*

or, putting for these areas their geometrical values,

$$DD' \times \tfrac{1}{2}AC > BAB' \times \tfrac{1}{2}AC;$$

suppressing the common factor ½AC, there remains

$$DD' > BAB',$$

and taking the halves, we have

$$DA > \text{arc } BA.$$

40. It follows from this, that the limit of the ratio of the sine to the arc is unity; for since, when the arc *h*, represented by AB, becomes nothing, the sine coincides with the tangent; much more does the sine coincide with the arc, which lies between the tangent and sine; and, consequently, we have, in the case of the limit, $\dfrac{\sin h}{\text{arc } h}$ or rather, $\dfrac{\sin h}{h} = 1.$

Boucharlat, Blakelock, 1828, pp. 24-5.

The ambiguity that surrounds the notion of limit allows us to juggle away[19] the passage from *h* tends to *0* to *h=0*. Conversely, we can find a confirmation of the

19. Jacques Derrida has devoted a chapter of his 1993 book *Specters of Marx the state of the debt, the work of mourning and the New international* to Marx' use of the French words " escamotage " or " escamoter ". Marx used several times these words that can be found in the German Ideology, written in 1841-1846. (See especially the chapter Sankt Max). Derrida's chapter is curiously entitled "The apparition of the unapparent". It includes a note by the translator stipulating that The English translation of The German Ideology retains "conjuring trick" for Marx and Engels' term Eskamotage. We will do so as well for obvious reasons, although the French word "escamotage" could also be translated by "dodge," "evasion," "filching," "pinching," and so forth". To be more specific, we can consider the definition to be found in the Dictionnaire de Trévoux. The original word seems to be "escamote" which is the name of a little cork ball which is taken subtlety between fingers. Hence "escamoter is the verb used by players of cups and ball in taking the cork ball (escamote) subtlety between fingers to realize some tricks. It means also: to make something disappear by a legerdemain unnoticed. Trévoux adds that "escamoter" means also, to steal subtlety and deftly. Marx uses repeatedly these words but nowhere more often than his mathematical manuscripts to explain how mathematicians obtain the derivative eliminating the infinitesimals. The English translation of the MMM uses various expressions to translate "escamoter" (to juggle away, conjuring tricks, etc.).

role played by ambiguity in the definition of the limit by considering the precautions Cauchy uses to proceed with the same demonstration:

The limits, which the 2 variable expressions $\frac{\sin\alpha}{\alpha}$, $(1+\alpha)^{1/\alpha}$ are indefinitely approaching, while α converges towards zero, present themselves in indeterminate forms, $\frac{0}{0}$, $1^{\pm\infty}$ and yet both limits have fixed values that can be calculated as follows.

Obviously, for very-small numerical values of α: $\frac{\sin\alpha}{\sin\alpha} > \frac{\sin\alpha}{\alpha} > \frac{\sin\alpha}{\tan\alpha}$

Therefore, the ratio $\frac{\sin\alpha}{\alpha}$ always lies between the two quantities $\frac{\sin\alpha}{\sin\alpha} = 1$ and $\frac{\sin}{\tan}$ = cos α, the first of which serves as a limit to the second, will itself have the unit as limit". (Cauchy, pp. 1-2).

It is striking to note that the precautions deployed here by Cauchy seem to have totally escaped H. Laurent, who adopts unchanged the first three paragraphs of Boucharlat, as well as the paragraph devoted to the limit of $\frac{\sin h}{h}$.

IV) Derivatives and differentials

It is in his § 4 that Boucharlat introduces the concepts of derivative and differential. This paragraph is reproduced below in its entirety from the 1826 edition (left-hand column) and from the revised edition by H. Laurent (right-hand column).

4. Since, on the hypothesis of h =0, the increment of y becomes also 0, $\frac{y'-y}{h}$ is reduced to $\frac{0}{0}$ and consequently the equation (2) becomes $\frac{0}{0} = 3x^2$ ...(3) This equation involves in it nothing absurd, for from Algebra we know that $\frac{0}{0}$ may represent every sort of quantity; besides which it will be easily seen, that since by dividing the two terms of a fraction by the same number the fraction is not altered in value, it follows that the smallness of the terms of a fraction does not at all affect its value, and that, consequently, it may remain the same when its terms are diminished to the last degree, that is to say, when they become each of them 0.	4. It is often necessary in Analysis to find the limit of the ratio of the increase that a function takes to the corresponding increase h of its variable: we give it the name of derivative or differential coefficient. Thus, the derivative of a function is the limit of the ratio of the increase that the function takes to the corresponding increase of its variable, when the latter tends towards 0. We represent the derivative of a function y of x by the notation $\frac{dy}{dx}$, an expressive notation which reminds us that y is the function, that x is the variable, and that we have taken a ratio of increases relative to each of these quantities.

The fraction $\frac{0}{0}$ which appears in the equation (3), is a symbol which has expressed the ratio of the increment of the function to that of the variable: since this symbol retains no trace of that variable, we will represent it by $\frac{dy}{dx}$; and then $\frac{dy}{dx}$ will remind us that the function was y and the variable x; but dy and dx will be no less evanescent quantities, and we shall have $$\frac{dy}{dx} = 3x^2. \qquad ...(4)$$ $\frac{dy}{dx}$ or rather its value $3x^2$ is the differential coefficient of the function y. We may observe that $\frac{dy}{dx}$ being the symbol which represents the limit $3x^2$, [as is shown by equation (4)], dx ought properly to be always placed under dy. In order, however, to facilitate operations in algebra, we may for a time clear equation (4) of its denominator, and we shall have $dy = 3x^2\,dx$. The expression $dy = 3x^2\,dx$ is what we call the differential of the function y. J.L. Boucharlat, § 4, pp. 3-4.	Note that $\frac{dy}{dx}$ being the sign representing the limit $3x^2$, dx ought properly to be always placed under dy.. However, in order to facilitate operations in algebra, we may for a timedisregard and clear equation (4) of the denominator dx and we shall have $dy = 3x^2\,dx$. This expression $dy = 3x^2\,dx$ or dy is what is called the differential of the function y; dx is then arbitrary and dy is nothing else than the product of dx by the derivative of the function. The differential of a function can therefore be defined as follows: *The differential of a function is the product of the derivative of the function by an arbitrary quantity called dx.* It goes without saying that if we take $dx = 1$, dy is reduced to the very derivative of y. Boucharlat-Laurent, § 4.

Here we can note a revision by Laurent which rules out what was more obscure in Boucharlat's synthesis between the Lagrange method and that of limits: it is clear here that, for Boucharlat, h effectively becomes null when it decreases. On the contrary, Laurent, without explicitly introducing limits in the sense of Cauchy, gives a formulation that is not incompatible with them.

This passage from Boucharlat seems to have drawn Marx's attention to the nature of the differential, and he cites it repeatedly. Indeed, he was particularly shocked by the way in which the separation of numerator and denominator is presented in the ratio $\frac{dy}{dx}$. He writes: "Thus, in order to facilitate algebraic operations" one introduces a false formula which one baptises the "differential", (Marx, 1983:25).

It is this inadequacy that will lead Marx to specify his own conception of the differential calculus. To do this, he will carefully compare the calculation of the derivative of a function of a single independent variable in two cases:

a) The simple case : $y = f(x)$

b) The product of two functions : $y = u*z$ with $u = f(x)$ and $z = \Phi(x)$

Calculating the derivative of $f(x)$, we get : $\frac{dy}{dx} = f'(x)$.

Now in this expression, $\frac{dy}{dx}$ is only a symbol, a sign that represents $\frac{0}{0}$ and Marx uses images: "ceremonial uniform", "Sunday dress". At the same time, he notes the absence in the right-hand side of the equation of symbols proper to the differential calculus indicating the operations to be carried out.

On the contrary, in the second case, Marx presents the differentiation in the following way:

$$y' = u' * z' z$$

$$y' - y = u' * z' - uz = z' (u' - u) + u (z' - z)$$

$$\frac{y' - y}{x' - x} \text{ or } \frac{\Delta y}{\Delta x} = z'\frac{u' - u}{x' - x} + u\frac{z' - z}{x' - x} = z'\frac{\Delta u}{\Delta x} + u\frac{\Delta z}{\Delta x}$$

Then, putting on the right-hand side $x' = x$, Marx draws from it: $\frac{dy}{dx} = z\frac{du}{dx} + u \frac{dz}{dx}$ or by elminating the denominator: dy or $d(uz) = z\ du + u\ dz$

In analysing the legitimacy of these operations, Marx insists on the difference in nature between $\frac{dy}{dx}$ which is the symbolic expression of $f'(x)$ in both cases and the $\frac{du}{dx}$ and $\frac{dz}{dx}$. These thus appear as "symbolic differential coefficients without real differential coefficients, i.e. without corresponding equivalent 'derivatives'. The symbolic coefficient thus becomes an independent starting point, the real equivalent of which is to be found first. [...] Whereas, initially, the differential coefficient is the expression of the derivative already carried out, it now plays the role of symbols of differentiation operations to be carried out." For Marx, this "reversal of the method" marks the birth of the differential calculus "which then operates independently on its own ground. For its starting points $\frac{du}{dx}, \frac{dz}{dx}$ belong only to it and are mathematical quantities characteristic of it. (Marx, 1983, pp. 20-21).

Marx then examines the separation between dx and dy and the definition of the differential, always distinguishing the same two kinds of function of x. He states, first of all, that in the first case, separation is legitimate when the substitution of dx for $(x' - x)$ causes the latter to lose "the form that manifests it as a vanished difference of the independent variable and likewise for dy [...]. Regardless of where dx is now located, such a change of location leaves the relationship between dy

and dx unchanged; $dy = f'(x)dx$ appears to us as another form of $\frac{dy}{dx} = f'(x)$ and is always likely to be written in this form". In the second case, he goes on to say, "In the second case, we saw that the differential coefficients were transformed into" independent starting points "or" symbols of operations "and likewise," symbolic equations are transformed into equations of symbolic operations. Thus, we have a double basis for treating the differential $dy = f'(x)\,dx$ as a symbolic equation of operation. [...] If the differential operation on $f(x)$ designated by $df(x)$ is carried out, the result will be $dy = f'(x)\,dx$ and this gives finally $\frac{dy}{dx} = f'(x)$. And Marx concludes that "it is only when the differential functions as a starting point of the calculation that the reversal of the algebraic differentiation method is completed and that the differential calculus itself would then appear a unique, specific method of calculating with variable quantities". (Marx, 1983, p. 26).

This 1881 text follows various drafts and reflects the conception of the differential calculus he had arrived at. It was the conception of the differential that attracted the attention of the commentators, but it is interesting to note that for Marx, the differential calculus is really founded from the moment the differential is conceived as a symbol of operation. Also before emphasizing the similarity between this conception and that of other mathematicians, it should be noted that this is precisely the approach adopted by Cauchy in his 1829 book. Cauchy first defines and studies the limits in "Preliminaries", then he begins his "1st Lesson" on "the object of the differential calculus" with a provisional definition of the differential:

> $x, y, z, \ldots$ being variables "determined by one or more equations, we call the differential of $x, y, z\ldots$ and we indicate it by means of the characteristic letter d in the notations: dx, dy, dz quantities whose ratios are equivalent to the final ratios of the infinitely small increases in these variables. The purpose of the differential calculus is to determine the ratios of the differentials $dx, dy, dz, \ldots$ when we know the relationships that exist between the variables, $x, y, z, \ldots$; [...]
>
> « To fix ideas », Cauchy then studies the case: $y = f(x)$.
>
> If the function $f(x)$ remains continuous between two given limits of the variable x and if a value between the two limits of the variable x is assigned to this variable, an infinitely small increase, attributed to the variable, will produce an infinitely small increase in the function itself. Therefore, if then $x = i$, the two terms of the ratio of the differences
>
> $$\frac{dy}{dx} = \frac{f(x+i) - f(x)}{i}$$
>
> will be infinitely small quantities. But while these two terms will indefinitely and simultaneously approach the zero limit, the ratio itself will be able to converge towards another limit, either positive or negative, which will be the ratio of the infinitely small differences Δy and Δx. This limit, or final value, when it exists, has a specific value for each particular value of x; but it varies with x. [...] The form of the new function that will serve as a limit to the relationship will depend

> on the form of the proposed function $y = f(x)$. To indicate this dependency, the new function is given the name of the derived function, [...]
>
> Having said that, the differentials dx, dy of the independent variable x and the function $y = f(x)$ will be the quantities so chosen that their ratio $\frac{dy}{dx}$ coincides with the last reason of the infinitely small quantities Δy, Δx i.e. with the limit $y' = f'(x)$ of the ratio $\frac{\Delta y}{\Delta x}$. These differentials will therefore be linked by the equation $\frac{dy}{dx} = y'$ or $dy = y'dx$ which can also be presented in one of the following forms : $\frac{df(x)}{dx} = f'(x)$ $df(x) = f'(x)\,dx$ (Cauchy, 1829, pp. 17-8).

Cauchy's approach is a good illustration of the reversal in the differential calculus that Marx had anticipated. The question then arises as to whether Marx was aware of Cauchy's works or works of the same type. There is no evidence that he had read Cauchy, yet one might think that Marx was at least aware of his work. In fact, on a list of books that Marx proposed to consult, one can find the *Leçons de calcul différentiel et de calcul integral, written according to the methods and published or unpublished works of M.A.-L Cauchy* by Abbé Moigno (Paris, Bachelier, 1840). But Abbé Moigno's fidelity to Cauchy is marred by notable elisions: first of all, we do not find in Moigno the care taken by Cauchy to define in his "Préliminaires" either infinitesimals or the limit of a variable. Moigno defines the former in a vague way, and for the limit he merely indicates that one "usually calls the limit of a function, the value towards which it converges when the variable on which it depends itself converges towards a given value". (Moigno, 1840, p. 3). Moigno then explicitly departs from Cauchy on the question of the definition of the differential. He writes in his "Introduction":

> In recent years, Mr. Cauchy believed that he had to give a direct, immediate definition of the differential, independent of the consideration of derived functions. Similarly to the approach of Maclaurin and D' Alembert, he calls differentials quantities whose ratios are equivalent to the final ratios of the increases that variables can take simultaneously. Starting from this definition one can calculate the differentials without going through the derivatives [...] but I did not want to take it as a starting point [...] I would have feared, by using it too soon to throw some obscurities on the principles of the Differential Calculus that it was so important to clarify". (Moigno, pp. XX-XXI)

So when Marx writes: "I remark further that from the early part of the eighteenth century, right down to the present day, apart from a few rare exceptions, the general task of the differential calculus has usually been formulated as follows : to find the real equivalent of the symbolic differential coefficient" (Marx, 1983, pp. 22), perhaps he counts Cauchy in the exceptions as a result of his reading of Moigno.

V) Various concepts of differentials

In 1934, Valery Ivanovich Glivenko (Kiev, 1896 - Moscow, 1940) reported the resemblance between the concept of the differential in Marx and in Hadamard. Fréchet studied different conceptions of the differential in an article celebrating Hadamard's achievements[20]. It can be used to highlight the inaccuracies or even errors into which H. Laurent has fallen, due to the lack of a correct definition. Fréchet writes:

> Assuming that the definition of the differential of a function of a variable is well known, M Hadamard proposed to define the differential of a function of several variables, for example, $f(x, y)$, as the classic expression
>
> $$(1)\ dz = \frac{\partial z}{\partial x} dx + \frac{\partial z}{\partial y} dy$$
>
> but for him, this expression is only a symbol of operations.
>
> Fréchet continues: "what does the equality (1) mean? That if x, y, and hence $z = f(x, y)$ are expressed as a function of an auxiliary variable u, we have whatever these expressions are
>
> $$(2)\ \frac{\partial z}{\partial u} = \frac{\delta z}{\delta x}\frac{dx}{du} + \frac{\delta z}{\delta y}\frac{dy}{du}$$
>
> This is the only meaning of equality (1). Since equality (2) takes place regardless of the variable by which the other two variables are expressed, the reference to "u" is deleted. (M. Fréchet, 1937, pp. 233-4).

Fréchet notes that "Hadamard's definition emphasizes the operative properties of the differential. Other definitions rather highlight its property of being a kind of principal component of the increase of the function". (Fréchet, 1937, p. 235). We can justly find an example of such conceptions in Laurent's *Treatise of Analysis*. H. Laurent does not clearly distinguish the two concepts of the differential. In fact, in his Treatise, after a Chapter 3 devoted to derivatives, he deals in turn in chapters 5 and 6 with the differentials of functions of one variable and those of the functions of several variables. In Chapter 5, after studying the "various orders of infinitesimals", he introduces the differentials in the following way: "It is advantageous to substitute $f'(x)\ \Delta x$ to $\Delta f(x)$ in such a way that one has expressed the need to give a name to this quantity $f'(x)\ \Delta x$; it is called the differential of $f(x)$". He then takes the case where $f(x) = x$, $\Delta x = dx$ and writes: $dy = f'(x)\ dx$; (H. Laurent, 1881, pp. 122-3). On the other hand, in Chapter 6, he seems to agree with an operative definition in paragraph I "on the calculation of symbolic expressions": "We have already seen that an operation symbol can be represented by a letter. Thus the operation which aims to take the derivative of φ and multiply it by h is represented by $d\varphi$. (ibidem, p. 133). But finally, the idea that differentiation is the main part of the infinitely small Δf seems to be much more characteristic than Laurent's conception, as can

20. M. Fréchet, "Sur la notion de différentielle", *Journal de Mathématiques Pures et Appliqués*, July-September 1937.

be seen precisely in the presentation of the derivative of a compound function. It assumes, for this reason, that "in function $f(x, y, z)$, the variables x, y, z cease to be independent, and that y and z are functions of x; according to the theorem of the composite functions, the derivative of f relative to x, which we will call the total derivative relative to x, will be given by the formula [...] $\frac{df}{dx} = \frac{\delta f}{\delta x} + \frac{\delta f}{\delta y}\frac{dy}{dx} + \frac{\delta f}{\delta z}\frac{dz}{dx}$ and he concludes a detailed discussion of the different differential notations with the remark: "It is almost useless to add that this is an infinitely small increase and that the terms of the second order are neglected" (ibidem, p. 141).

One can see in the use of this definition of the differential of a function of several variables the errors to which it can lead. H. Laurent believes that it is indeed possible to establish the "fundamental theorem" :

> The increase that a function of several variables undergoes when one gives increments of the same order to its variables is, except for the infinitesimals of higher order, equal to its differential, and consequently to a sufficient degree of accuracy, the increase of a function can be replaced by its differential without changing the result. (H. Laurent, p. 148).

However, following the work of various mathematicians (Otto Stolz and W.H. Young) of the late 19th and early 20th centuries, as M. Fréchet writes[21]

> it is easy to see that even for very simple $f(x, y)$ functions, the ratio $\frac{\Delta f}{df}$, *i.e.* $\frac{f(x+\Delta x, y+\Delta y)}{f'_x \Delta x + f'_y \Delta y}$ does not necessarily tend towards unity in any way when Δx and Δy tend towards zero. (M. Fréchet, 1937, p. 235).

Engels used ironic words about mathematicians in general after becoming familiar with the critical reading of the history of differential calculus that Marx presented to him[22]. They were probably not justified for all mathematicians. Similarly an overall assessment of Marx's mathematical work would require a careful analysis of his history of Taylor's and Maclaurin's theorems and their role in the differential calculus. But it seems legitimate, as of now, to bring Marx's concerns closer to the modern presentation of the differential. Let us recall that:

a) Taylor's formula approximates the difference $f(x + h) - f(x)$ by $f'(x)h$. We call the differential of $df(x)$ in x, the application noted $df(x)$:
$R \to R$
$h \to df(x)\ (h) = f'(x)h$
$df(x)$ is linear.

21. Mr Fréchet quotes Otto Stolz and W.H. Young. He probably has in mind O. Stolz: "B. Bolzano's Bedeutung in der Geschichte der Infinitesimalrechnung", *Mathematischen Annalen*, XVIII, 1881, 255-279. Bourbaki's *Elements* mentions various works on the integral calculus by one W.H. Young published in 1911.
22. "The thing is as clear as daylight, so that we cannot wonder enough at how stubbornly mathematicians insist on mystifying it" (Letter from Engels to Marx, August 18th, 1881).

b) the differential of $df(x)$ is then defined as the application noted df of R in the set L(R) of the linear applications of R in R.

$df : R \longrightarrow L(R)$

$x \longrightarrow df(x)$

However, what explains the difficulties encountered by mathematicians of the 19th century (and after) and the confusion in this area is that:

$$L(R) = R.$$

Marx saw these difficulties and his concern for dialectics led him to carefully distinguish the process (the operation of differentiation) from the result (the derivative), but, not having at his disposal the modern tools of set theory, he was led to propose different formulations and distinctions (e.g. the symbols of operation). It appears, however, that Marx's interest in the transformations brought about by the development of different approaches to the differential calculus suggests that the genetic and dialectic method (the history of concepts) deserves attention.

Alain Alcouffe

Autumn 1986.

University of Social Sciences of Toulouse.

The author would like to thank M.J.C. Yakoubsohn, assistant professor of mathematics at the University Paul Sabatier in Toulouse for the help provided both in the preparation of the French edition of Marx' Mathematical Manuscripts and in the reading of Boucharlat or Laurent, as well as for the enlightenment given on the modern conception of the differential.

The English translation benefited from Julian Wells' revision and editing.

The author remains, of course, solely responsible for any errors that may remain.

The original French version of this paper :

Alcouffe, Alain (1987), "Le calcul différentiel, les mathématiciens et les économistes au XIX° - K. Marx et H. Laurent, lecteurs de J.L. Boucharlat" *Sciences et techniques en perspectives*, 1987, 13: 47-69. It is available at :

https://hal.archives-ouvertes.fr/hal-01312597/

___

The Author :Alain Alcouffe (1945-), Université Toulouse 1 Capitole, Economics, Emeritus: <http://alain.alcouffe.free.fr/>.

He edited *Les manuscrits mathématiques de Marx* (1985), Paris: Economica.

Part Three

Plural Mathematics

Plural Mathematics: Before and After Marx

Mathematics and its history in retrospective

Adolph-Andrei Pavlovich Yushkevich

The present paper is a revised and supplemented version of a report read at the All-Union Symposium on "The Regularities and Modern Tendencies of the Development of Mathematics", held in September 1985, at Obninsk. The report dealt with the changes that have occurred in our outlook about the development of mathematics from the ancient times to the modern period, as a result of the recent historico-scientific investigations. While preparing this paper I thought that it would be advisable to provide a prefatory retrospective, as to how the past of mathematics is viewed to-day: it would be a short retrospective review of the historico-mathematical investigations themselves, The literature used for preparing this paper is too vast, and the list that follows the text of the paper contains references to only a few books or papers, indicated against their corresponding numbers, more often only the name of the author of a work and the year of its publication has been mentioned.

Understandably, here we shall be dealing with only some of the changes that have taken place in our ideas about the development of mathematics at the different stages of its formation as a science. Nowadays, sometimes we come across the view that the mathematics of ancient Egypt, or Babylon, or China was yet to become a science and it became one only in ancient Greece. However, the historians of science are yet to agree as to which fields of learning may be called a science and which may not be. The present author is merely of the opinion that the aforementioned view is not sufficiently substantiated, and on this more later.

A Historiographic Retrospective. Historiography of mathematics dates back to antiquity. One finds its odd elements in the works of Plato and Aristotle, whose pupil Eudemus of Rhodes (incidentally, not a mathematician) was the first to author a treatise on the history of geometry. Afterwards, individual scholars did turn their attention to the history of mathematics; but their work have long lost all significance. The study of the history of science was highly valued by such leaders of scientific and philosophical thought as F. Bacon and G.W. Leibniz; and in the days of Enlightenment, its leading ideologists saw the motive force of progress in the growth and spread of knowledge, wherein mathematics and mechanics (inclusive of celestial mechanics) became the leading sciences. The first fundamental work on the history of these disciplines and of some parts physics—"The History of Mathematics" by the Parisian academician Jean-Étienne Montucla—was published in this period. Its first two-volume edition appeared in 1758 and, the second, much enlarged four-volume edition came out during the period 1792-1802, only after the death of its author. This work was carried out through to the end by the astronomer Joseph Jérôme Lefrançois de Lalande and mathematician Sylvestre-François Lacroix [1]. This book was a great work of its time; in spite of the then unavoidable gaps, inexactitude and the dated methodology, the modern reader will find interesting information in it, which however, should be used with circumspection. Almost simultaneously with the publication of the second edition of Montucla's work, came out a two-volume general history of mathematics by another academician of Paris—Charles Bossut (1st ed., 1802), and a four-volume history of the physico-mathematical sciences by a professor of the Gottingen University Abraham Gotthelf Kästner (1796-1800).

The scope of the investigations into the history of mathematics was continuously widened during the 19th century .The study of the primary sources of mathematics of the people of medieval Orient began, the first translations of the works of the Arab and the Indian mathematicians appeared, critical editions of the works of the Greek authors Euclid, Archimedes, Apollonius and of the others were prepared, similar editions of a number of classics of the modern times were also begun, sometimes to be concluded only in the 20th century—these included the works of the mathematicians from R. Descartes and P. Fermat to A. Cauchy, B. Riemann and K. Weierstrass. The work in this direction has continued with greater intensity in the present century. Thus, the old plan of publishing the complete collected works of L. Euler (which began in 1911 and is nearing its completion only now) is being realised. The works of C.F. Gauss, N.I. Lobachevsky, H. Grassmann, P.L. Chebyshev, Jules Henri Poincaré, D. Hilbert and of others have either been published in full or in selections. [For more detailed information see: the books by G. Loria [2] and K.O. May [3], which mention the classics published upto 1946 and 1973 respectively]. During the last decades of the 19th century M. Cantor, Baldassarre Boncompagni, Gustaf Hjalmar Eneström and V.V. Bobynin took the initiative to start the first journals of history of mathematics [see: 2] and, the first courses on this subject were introduced in some of the universities—this is a rare phenomenon even to-day. The literature on the history of mathematics grew and, the famous four-volume history of mathematics by M. Cantor was published during the period 1880-1907 [4]; however, the fourth volume of this book was written by a group of scholars under the editorship of this great historian of mathematics. The work of Cantor covers the period upto 1799. It is still an useful reference book, though in certain parts it has become entirely outdated; what is more, in it the development of mathematics is viewed only in itself, outside the framework of general history and, often not even in connection with the mathematical natural sciences. The two excellent books by Hieronymus Georg Zeuthen on the history of mathematics upto the beginning of the 18th century, at first published during the years 1893-1903, are of a different character: there the mathematical treatment of the subject-matter is much more deep in comparison to Cantor's work; it is true though that they contain less of the details. Both of them have been translated into Russian [5]. In this period the interest in history of mathematics grew considerably among the mathematicians themselves, especially in the history of those disciplines in which they specialized. Hence the works on history of geometry by Michel Chasles (1837), those of Isaac Todhunter (variational calculus, 1861; theory of probability, 1865), Alfred Enneper (elliptical functions, 1876), I. Yu. Timchenko (theory of analytical functions, 1899) and others. The aforementioned Zeuthen was an outstanding specialist in algebraic geometry and a person of broad outlook.

Towards the end of the 19th and the beginning of the 20th centuries, the growth in the interest about history of mathematics was considerably promoted by the great German scientist and one of the initiators of the movement for the reform of mathematics teaching in the secondary schools. F. Klein. A three volume monograph on elementary mathematics, treated from the point of view of higher mathematics emerged out of his lectures read to the teachers of Göttingen University. First published in 1903, this book is saturated with historical materials. Its Russian translation saw two editions [6]. The history of elementary mathematics by the German pedagogue and scholar J. Tropfke [7], first published in a two-volume edition (1902-1903) and then extended upto seven volumes (1921-1924), was mainly intended for teachers. After a long gap K. Vogel and his collaborators decided to prepare a new edition of "Tropfke" and in 1980 a structurally quite excellent edition was brought out; this book on the history of arithmetic and algebra fully corresponds to the contemporary state of our knowledge on the subjects [8]. The subsequent new "Tropfke" (geometry) has not followed, however, owing to the demise of K. Vogel. Yet another widely known history of elementary mathematics was authored by the American scholar F. Cajorie; it was published in 1896. Its Russian Translation (1910) is accompanied by highly valuable supplements from the translator I. Yu. Timchenko [9].

Like many other mathematicians F. Klein too attached a great cognitive significance to the history of mathematics. He took the initiative to include innumerable historical information into the famous German six-volume encyclopaedia of the mathematical sciences (1898-1934). Klein wrote in the preface of this practically nearly all-embracing collective work [10], that in it there should not only be a concise and generalised presentation of the modern condition of the mathematical sciences with their applications in the other sciences and in technology, but also a description of the evolution of the mathematical methods from the beginning of the 19th century should be provided with the help of carefully selected records and reference literature. There exists an incomplete French version of this encyclopaedia (1904-1914). One finds a fuller representation of the history of mathematics in the Italian encyclopaedia of elementary mathematics—published in 3 volumes and 6 books under the guidance of Luigi Berzolari, Giulio Vivanti and Duilio Gigli, in the years 1932-1950 [11]. Here, apart from the information contained in the main text on the history of elementary mathematics, there are independent sections on the main trends of modern mathematics and on the questions of didactics; and they follow the section on elementary mathematics. This "Encyclopaedia" is very rich and is within the reach of the students of the first years of any university. However, unfortunately, it is not well known beyond the borders of Italy. Running ahead, let me add here: historico-mathematical essays or sections of essays occupy a prominent place in all the three editions of the Great Soviet Encyclopedia, thanks to the unfailing directions of the editors of the GSE, in particular of V.F. Kagan and A.N. Kolmogorov.

The ever growing interest in the history of mathematics and the recognition of its status as an independent and important section of the entire system of the mathematical sciences, in the 19th and 20th centuries, may be illustrated with the help of many facts; from among them I shall adduce only three. Two of them belong to the very beginning of the 20th century; they are, both related to the Second International Congress of Mathematicians held at Paris, in the summer of 1902.

The first incident is D. Hilbert's famous paper on the "Problems of Mathematics", read in this Congress on the 8th of August. In this paper 23 real problems from various areas were posed; they exerted a strong stimulating influence upon the subsequent development of mathematics. Hilbert's problems are usually viewed from the mathematical angle of vision, and this is understandable. However, there is another side to this affair: Hilbert's judgements on the perspectives for the development of mathematics and the sorting out of its especially real problems, are based on a deep going analysis of its previous development. In his own words: "History teaches us that the sciences develop uninterruptedly. We know that every age has its own problems, which are either solved or are set aside as fruitless and, substituted by new ones, in the next epoch. In order to conceive the possible character of development of mathematical knowledge in the near future, we must take a look at those questions which still remain open, and survey those problems, which have been posed by modern science, whose solution we expect in the future. It seems to me that such a survey of the problems is especially contemporaneous today, at the dawn of a new century" [12]. Hilbert was an eminent mathematician of the end of the 19th and beginning of the 20th centuries, like H. Poincaré, and it is thus that he posed the question of the study of the past of mathematics, in the interest of a creative prognostication of the perspective for mathematics.

The second noteworthy fact was the placing of M. Cantor's paper "On the Historiography of Mathematics" in the same Congress. It was one of the 4 plenary reports. Cantor began his survey with Montucla's work. There were 6 sections in the Congress. The sessions of the section on history and bibliography were held jointly with the section on leaching and methodology; these sessions were presided over by G. Cantor. The plenary report of V. Volterra on the life and work of three great Italian mathematicians—E. Betti, F. Brioschi and F. Casorati—was also historico-biographical in character. In all the subsequent congresses since then, there was always a section on the history of mathematics.

As the third example, mention may be made of the five-volume Soviet "Mathematical Encyclopaedia" (1973-1985). It contains innumerable historical information and references as a matter of course, though there are no historico-mathematical articles proper.

Omitting the events of the historico-mathematical life till the end of the first world war, which disturbed the normal course of scientific progress, let us turn to the last half a century, marked by ever growing activisation of the historico-mathematical investigations, wherein a special mention must be made of the last 20-30 years. [One must mention, however, that, namely, in the years 1914-1919, F. Klein read his remarkable lectures on the development of mathematics in the 19th century, to a small circle of listeners, who gathered in his flat. These lectures were later on prepared for publications by R. Courant and O. Neugebauer; they were published in 1926, a year after Klein's death. A Russian translation of the first, historico-mathematical, part of these lectures was published in 1937 [13]. The second part is devoted to physics at the end of the 19th and beginning of the 20th centuries and to its mathematical apparatus; it contains short historico-scientific digressions, but they play a subordinate and insignificant role in it.] In this period many socio-historical, general cultural, ideological and scientific-organizational factors were in operation. We shall not list all of them, nor enunciate them in terms of their importance and shall be mentioning only some of them.

First of all, history of mathematics, like that of the other sciences, is organisationally constituted, with material support both at the international level and at that of the individual states. In 1929 the International Academy of History of Sciences was created and the first International Congress of the History of Sciences was held, at the initiative of a group of leading scientists from many countries. At present this Academy has nearly 230 full and corresponding members in its rolls; they are from many countries (26 of them are from the Soviet Union) (from the erstwhile USSR—Tr.). This Academy publishes its journal since 1948, and since 1968 has begun awarding a prize for outstanding scientific excellence, in the name of the great French historian of science A. Koyré. After the founding of the UNESCO, an International Union of the Historians of the Sciences was organised as its subsidiary, just like the International Union of Mathematicians, Specialists in Mechanics etc., all of which consisted of separate national unions. The Soviet Association of the Historians of Science and Technology was established in 1957. There is no formal, juridical connection between the International Academy and the International Union of History of Science, but in reality the members of the Academy occupy all the leading posts in the Union; and there are Soviet scholars among them. An important job of the Union and of its national sections is to organise international congresses (from 1929 to 1985 there have been 17 of them), national conferences, and symposiums of a general, as well as of a specific, character—dedicated to individual disciplines, problems, jubilee celebrations etc.

The most important pre-condition for speeding up the progress of investigations on the history of mathematics and of the other sciences, is the preparation of better workers with Suitable specializations, in the institutions of higher learning and, the establishment of institutes of history of science. In different countries this has been done in different ways: great success has been achieved in this respect in the FRG, GDR (now Germany—Tr.), USSR (now former USSR—Tr.), France and, of late in China; the USA has not been mentioned here, since there the preparation of workers and the organization of researches have their own specificities, and there is no scope for dwelling upon them in the present paper. As an example, here we shall briefly narrate the state of affairs in the USSR, mainly in the two centres at Moscow—one in the University, and another in the Academy.

Before 1917, (aforementioned) V.V. Bobynin taught an optional course on the history of mathematics, in the Moscow University, for quite some time. In the 30s, this course was renewed and later on made compulsory. A regular scientific seminar on the subject began to function since 1933—it obtained all-union recognition; research studentship on the history of mathematics was introduced, doctoral and post-doctoral work began to be defended in this specialization. After the second world war a special section was created for the histories of mathematics and mechanics, students' seminars were organised, and graduation theses were introduced in these disciplines. All this brought forth tangible results. This section is connected with the other kindered organizations in Leningrad, Kiev, Tashkent etc., as well as with the corresponding centre in the Academy of Sciences, USSR.

An interest in the history of mathematics existed in the Academy of Sciences of the USSR, since long. In the first years after the October Revolution, the Academy at first ordered the printing of A.V. Vasiliev's book on the development of mathematics in Russia from the epoch of Euler to that of Chebyshev (1921), and then of V.A. Steklov's book on mathematics and its significance for mankind (1923)—a book saturated with historical material. Like V.A. Steklov, many other members of the Academy, namely, A.N. Krylov, V.I. Smirnov, S.I. Vavilov, T.P. Kravets, R.O. Kuzmin, N.G. Chebotaryov etc.—this list may be extended considerably—took an active interest in, and did study, the history of the physico-mathematical sciences. In the 30s and 40s certain measures were adopted, with the aim of imparting a more regular character to the investigations on the history of the sciences and technology, conducted in the Academy. The founding of the Institute of History of the Natural Sciences in 1945, and after its amalgamation with the Commission on the History of Technology in 1953, its transformation into the Institute of History of Natural Science and Technology (with a branch in Leningrad)—was of decisive significance. It is the biggest institution of its kind in the world. Here a highly qualified group of historians of mathematics works in close collaboration, not only with their fellow-workers in Moscow and in the other Republics of the Union, but also with many foreign scientific centres and individual scholars, above all with those from the GDR, FRG (now united Germany—Tr.)., China, USA, France, Czechoslovakia (now the separated Czech and Slovak Republics—Tr.) and, Switzerland. In this Institute too, a permanent scientific seminar on the history of mathematics works and, specialists—research students and doctoral candidates—are helped in their work.

The total number of the more or less active historians of mathematics today, is not known. According to a directory published in 1978 [14], at that time their number was nearly 1500, now it must be considerably more and, probably, about 2000. Here, among other things, one must have in view the fact, that now investigations in this field are being conducted, not only in those countries, where the corresponding tradition has long since been established, but also in those, where earlier there did not or almost did not exist a national contingent of historians of the sciences; such countries include: the many Arab states, Turkey, India, China, Japan, Canada, those in Central and South America, and of course those Republics of the USSR (now CIS—Tr.), which constituted—socially and culturally speaking—the backward periphery of the Russian empire, before the October revolution. One of the consequences of the global decolonization of the earlier possessions of the imperialist states, has been a rapid growth of interest in them, in their own cultural past, and in general, in history.

Investigations in this field grew almost every month. The need to publish them, in turn, replaced the periodicals, that had been discontinued since the beginning of the 20th century. New historico-scientific journals or series of occasional thematic collections appeared; after the second world war their number grew—and continues to grow considerably.

A list of such publications, keeping it limited on the one hand to those that are more specialized, and on the other—to the most well known, in the chronological order of their appearance, is as under: "Istoriko-matematicheskie issledovaniya" ["Historico-Mathematical Investigations"] (1948-), published in the Russian language (wherein the papers of foreign authors are printed in their Russian translations; however, these publications mainly contain papers of the Soviet scholars); "Archive for History of the Exact Sciences" (1960-), publishes papers in the main European languages, save Russian (this is connected with the purely external conditions of publication—in this journal the papers of the Soviet authors are printed in other languages); and finally, the organ of the Mathematical Commission of the International Union of the Historians of Science—"Historia Mathematica" (1974-), publishes papers in 10 European and Oriental languages, as well as information on scientific activities, reviews and bibliographical surveys. There exists no data about the publications on the history of mathematics, though one may get some idea about their number from the fact that the 30 issues of the "Istoriko-matematicheskie issledovaniya" (owing to purely external reasons they were not published during the years 1967-1977) contained 600 papers. Apart from the aforementioned publications, there are the collections published by the Section of History of Mathematics and Mechanics of the Moscow University, by the history of mathematics seminars of the Jules Henri Poincaré Institute of Nantes and Toulouse and, the new journals being published in India, Japan etc. All this has not only opened new opportunities for publication, but has also stimulated further investigations.

We have stated above the approximate number of the investigators in the field of history of mathematics, and have added that their number is growing. We must mention in this connection the fact that simultaneously the mathematical preparation of these investigators has also improved—and this as a result of an improvement in the education of higher mathematics, determined by the rapid progress of the mathematical sciences themselves. The linguistic training of the historians has also improved: now original researches have become impossible or do become inferior in quality, without a solid knowledge of many languages.

The directory of historians of mathematics [14] contains the names of quite a few mathematicians—specialists in this or that branch of this science. Generally speaking, the relation of the mathematicians with the history of their own science has not been and is not uniform. It has lived through times of rise, as well as those of fall.

There are specialists, who are not at all interested in the history of science. On the other hand, one can count the names of dozens of mathematicians, among them there are those who are of the highest class, who are not only interested in the history of mathematics, but are active in this field (in particular, they add historical sections to their manuals), and they also organisationally co-operate with the progress of history of mathematics, at the level of organization of science. It is impossible to list the names of all such scholars. It is enough to name from among those of the older generation: P.S. Alexandrov, A.D. Alexandrov, B.L. van der Waerden, A. Weil, H. Weyl, J. Dieudonné, A.N. Kolmogorov, V.I. Smirnov, D.J. Struik; and from among the younger ones—J. Dhombres, H. Koch, Yu. I. Manin, A.N. Parshin, A.D. Soloviev, V.M. Tikhomirov and C. Houzel. Of course this is a personal selection, and quite fortuitous at that, and the names of many a top mathematicians have been left out here—mathematicians who are systematically building bridges between mathematics and its history.

The heightening of the interest in history of mathematics among the mathematicians, especially among the scholars with a broad range, in the first half of the 20th century, had, in part, been conditioned by the crisis in the foundations of mathematics and the discussions generated by it, which drew the attention of many scholars to the historical retrospective. It was also influenced at a different level (in the first place, in our country), by the publication of the Russian translation of a part of the "Mathematical Manuscripts" of K. Marx, in 1933. [A greater part of Marx's Mathematical Manuscripts were published in 1968, though a complete edition of these manuscripts remains to be published.—Tr.] One of the consequences of the aforementioned discussions has been a tempestuous progress of mathematical logic and, the follow-up action still continues. During the last few decades, the "storms" in the development of informatics and of the adjacent fields, as well as a revolution of its kind in computational mathematics, has again drawn the attention of a number of specialists to history. Clearly, today, in principle new paths are being outlined for the development of mathematics and, one of the means of trying to understand the paths of its further development, is to turn to its retrospective.

Cooperation among the mathematicians—the historians and the specialists, has already become an imperative necessity in our time. It is necessary for both the groups, and it is already yielding good results. Perhaps, here priority should be accorded to: 1) the publication of the classics and 2) to writing generalised works on the history of mathematics of the modern and the recent times. It is enough, for instance, to mention the publication of the collected works of Euler, Gauss, Lobachevsky, Ostrogradsky, Riemann, Chebyshev, G. Cantor, Lyapunov and of Markov Sr. All these modern publications are accompanied by commentaries, without which it is difficult, and in any case less effective, to study the works of the said scholars. In this respect the recent editions of the classics of mathematics, are as a rule qualitatively superior to those of the 19th and early 20th centuries.

In respect of the works on the history of mathematics, one may mention the multi-volume Soviet history of mathematics from the ancient times to the beginning of the 20th century [15], the publication of which is still in progress, and the French history of mathematics of the 18th-19th centuries, wherein, in many sections, the 20th century too has been dwelt upon [16]. The books dealing with the history of mathematics in large geographical areas, as for instance, the collective works on the history of mathematics in our country, which cover the subject almost upto our time [17, 18], are of great value.

Jointly authored books, of the type just mentioned, are generally speaking preferable to the monographs produced by single authors: in our time one person cannot produce a balanced and thoroughgoing work on the history of mathematics from the ancient to the modern times. To be specific, the American mathematician M. Kline could not do it; his book [19] contains very interesting and competently written chapters, yet, in spite of its volume—containing 1248 pages—there are very substantial problems, related to important mathematical disciplines, for instance, regarding the theory of probabilities, and in respect of some regions, like China etc.

We have already mentioned two generalised works on the history of mathematics [15 and 16]. Now, a few words about the general orientation of the Soviet and the French collectives are in order. In the French case it was perhaps determined by the leader of the group. In the Soviet work, mathematics has been considered, not only at the level of its ideational development or self-development, but also as a social phenomenon, in its interconnections with the social requirements, with the other sciences, engineering, philosophy etc., briefly speaking, in the interconnections of the superstructure with the base (it is not for the present author—a member of the editorial and authorial collectives—to judge, how far this attempt has succeeded). In contrast, in the French work, attention has been concentrated, save in a few points of the introduction, upon the self-development of the ideas of the so-called "pure mathematics", which have almost exclusively been considered at the level of their immanental interconnections. [In the introduction of this work it has been said that "the most elementary concepts of modern mathematics" have been considered "in their historical contexts" and, in interconnection with their applications in the natural sciences. However, in the course of the work this declared objective has been very timidly realised.] It must be stressed, that in this work, generally speaking, one finds a very deep and substantial mathematical analysis of the historical process: almost all the authors are specialists in their respective fields of mathematics, who have painstakingly studied the essential literature on a given question, including many works of the mathematicians of our country (and this is not true of M. Kline's book).

What we have said about this book [16], also holds good for an earlier work of N. Bourbaki—a remake of the historical essays contained in the various volumes of their "Elements of Mathematics", which have been published since 1939; this re-writing was mainly done by A. Weil and J. Dieudonné [20]. It may be stressed here, at the same time, that in the recent years, one more often notices a greater awareness about the social factors related to the development of mathematics, also in the foreign literature on the history of mathematics. The widely popular "Concise History of Mathematics" by the American geometer D.J. Struik, which was first published in 1948 and has since been reissued many times in English and in numerous translations, including one in Russian [21]—happens to be a notable example of this kind of book. Unfortunately, the valuable supplement by I.B. Pogrebyssky, characterizing the mathematics of the first half of the 20th century—laid aside by the author—has been dropped, in the latest Russian edition of this book.

Along with the generalized works, the last half a century has also seen the publication of many original works, including such monographs, as have substantially deepened our knowledge, as well as of translations from the Oriental languages—less known to the circle of historians of mathematics of Europe and USA—into the languages of Europe. These works encompass a very large time span and geographical territory. In the short list that follows we shall indicate only the names of the authors and the years of publication.

One must begin this part of the historiographical survey, with V.V. Struve's edition of the Moscow-Egyptian-Papyrus (1930), which significantly augmented our knowledge of the mathematics of *ancient Egypt*—earlier this knowledge was almost exclusively based upon the so-called Rhind Papyrus (1877). Next, we must mention the publication of and investigations upon the cuneiform *Sumero-Babylonian* texts by O. Neugebauer (1934-1951), F. Thureau-Dangin (1938), E.M. Bruins (1957), A.A. Vaiman (1961) and others. In the field of ancient Greek mathematics one has to mention, at least the works of O. Becker (1933-). M.Ya. Vygodsky (1941), I.G. Bashmakova (1958-), van der Waerden (1950). A. Szabó (1955-) and Jean-Pierre Vernant (1962). The study of the mathematics of the *middle ages* has been conducted in a number of regional directions. E.I. Bereozkina translated almost the entirety of the so-called "Ten books" from *Chinese* into Russian; she also came out with a preliminary survey of her investigations, in a book published in 1980. The Japanese scholar Y. Mikami gave us the first sufficiently adequate description of the history of mathematics in *China and Japan*, in the English language (1913). His subsequent important papers are in Japanese and remain almost unknown in Europe, till date. In China proper, important investigations began later, first of all in the works of Li Yan and Qian Baocong (1935-1937); their main essays still remain to be translated in the European languages. At present a large group of Chinese and European specialists are working on this problematique, and many important discoveries have been made—which have often been described only in the Chinese literature. In the recent years, the original works of K. Chemla (of France), A.K. Volkov and those of the other young specialists have been published. Chronologically speaking, among the comprehensive works, first comes the volume devoted to mathematics in the multi-volume history of civilization in China, published in 1954; this joint work of J. Needham and Wang Ling contains a very rich bibliography, which is now, understandably, somewhat dated [see: the bibliographical survey in the just mentioned book by E.I. Bereozkina, which, naturally, does not contain any reference to the publications since 1980]. Yet another direction of research had the mathematics of *India* as its subject matter. The first stage of the investigations in this field has been summed up in a two-volume book by B. Dutta and A.N. Singh (1935-1938); subsequently, interesting investigations were conducted into the infinitesimal mathematics of India of the 15th-16th centuries by C.T. Rajagopal and T.V. Vedamurthy (1949-). Afterwards other Indian scholars, and to a lesser extent European scholars also conducted their investigations in this area, A.I. Volodarsky's book (in Russian) on the mathematics of medieval India came out in 1977. A great cycle of work has been conducted on the mathematics of the *Arab countries, Iran and Central Asia*. Here one must at least mention the names of C.P. Luckey (since 1938-), E.S. Kennedy (1947-), R. Rashed (1968-), J.G. Vernet (1952-), Julio Samsó (1966-) and, from among the Soviet scholars those of B.A. Rozenfeld (1951-), G.P. Matvievskaya (1961-), as well as their many colleagues and pupils. [In 1983 G.P. Matvievskaya and B.A. Rozenfeld published a bibliography of the literature on Arab mathematics and astronomy, in three volumes. It is a much more detailed bibliography, than the one published by H. Suter in 1900-1902]. Finally, the fourth direction in the study of the history of medieval mathematics: its study within the frame-work of the *European region*. Here, the aforementioned K. Vogel—engaged in the study of the development of elementary mathematics in medieval Europe (and Byzantium)—made great contributions. However, the numerous works on the higher mathematics of medieval Europe, are of special interest. Here significant contributions have been made by P. Duhem, V.P. Zubov (since 1947-), A.C. Crombie (1953), H.L. Crosby Jr. (1955), M. Clagett (1959-), C.A.Wilson (1960), H.L.L. Busard (1961-), J. Murdoch (1961), V.S. Shirokov (1978-) and others, who have continued, and introduced more clarity to the investigations of the pioneers. Compared to the earlier understanding of the subject, medieval mathematics and its role in the global progress of science now stand illumined in a completely new light.

The principal works on the mathematics of antiquity have in the main been summed up in B.L. van der Waerden's well known monograph (1950), and in the history of medieval mathematics, penned by the present author (1961); both of them have been translated in a number of languages; and in view of their years of publication, both appear somewhat outdated on a number of points, in the light of our present level of knowledge of the subject.

Neither the Arab countries, nor Europe knew of book printing with the help of moving types, in the middle ages—it began only in the middle of the 15th century—and, books were brought out in the manuscript form. That is why the historians of the sciences of this period are required to look for manuscripts and the search yields rich hauls. However, on the other questions too, history of mathematics is largely indebted to archival investigations: these are related to the works of Newton, Leibniz,

Euler, Cauchy, Bolzano, Ostrogradsky, Bunyakovsky, Chebyshev, Kovalevskaya, Weierstrass, Dedekind, Luzin and others. In the instances herein mentioned and in many other instances, the obtained archival materials were of great significance, not only for the exact dating of various discoveries or for solving the questions of disputed priorities but also for the discovery of hitherto unknown aspects of the creativity of the great scholars, of the activities of large scientific collectives, of the international scientific community, of the emergence of scientific contacts among individual scholars and among the institutions, in which they worked, etc. As an example one may mention the three volumes of L. Euler's correspondences, letters that he wrote to the Peterburg Academy from Berlin, in the years 1741-1765. In this period he was a foreign member of the Peterburg Academy and, a full member of the Berlin Academy (he returned to Peterburg in 1766, where he had earlier worked from 1727 to 1741).

The complete history of mathematics, like that of any other science, consists not only of the history of its ideas, but also of the history of the people who created that science, that of their collectives. In this connection one may mention the fact that specialists and historians of mathematics (as well as those of the other sciences) took part in compiling the famous 16 volume dictionary of scientific biographics, published under the editorship of Gillispie (1970-1980). Same is true of that big Italian biographical dictionary (1975), which contains less detailed biographies, as compared to the former, but is very rich in illustrations and provides a substantial survey of the development of the sciences since 1875 (more than 500 pages). Paucity of space forbids us to dwell upon the books about individual scholars, though they too are an inseparable part of the historico-scientific literature.

During the last 25 years a large number of monographs have been published on the history of individual disciplines; these are highly useful even for the specialist mathematicians. It is impossible to give a complete list of these monographs here and one is constrained to limit oneself to a few examples. All of them are of a very high scientific standard, though at times quite subjective in their evaluations of the role of individual discoveries or scholars. Such are some of the published monographs: on the history of the theory of numbers and algebra (Hans-Ludwig Wußing 1963; L. Nový 1973; A. Weil, 1983; B.L. van der Waerden, 1985; I.G. Bashmakova and E.I. Slavutin, 1985), on the development of set theory and theory of functions (F.A. Medvedev, 1965-1982; U. Cassina and M. Guillemot, 1983), on the history of the foundations of analysis from Euler to Riemann (I. Grattan-Guinnes, 1970; Christian Houzel, Jean-Louis Ovaert, Pierre Raymond, Jean-Jacques Sansuc, 1976; a book on R. Dedekind—P. Dugac, 1976; U. Bottazzini, 1981), on the history of the trigonometric serieses (A.B. Paplauskas, 1966), on the history of the theory of functions of the complex variable (S.E. Belozerov, 1962), on the history of differential equations and of functional analysis (C. Truesdell III—on the equations, 1960; V.A. Dobrovolsky, 1974; V.S. Sologub, 1975; J. Lützen, 1981), on computational mathematics and computing machines (H.H. Goldsteine, 1977; I.A. Apokin and L.E. Maistrov, 1974), on the theory of probabilities (L.E. Maistrov, 1967 and 1980), on non-Euclidean geometry (B.A. Rozenfeld, 1976), topology (Jean-Claude Pont, 1974), and logic (J.M. Bochenski—in English—1961; N.I. Styazhkin, 1964; T. Kotarbiński, 1965). It is an incomplete list, but even if it is supplemented with a few more names of hitherto unmentioned monographs, even then that would not encompass all the basic mathematical disciplines. Work in this direction is of first order importance, and it is still continuing. Here, owing to insufficiency of space, we shall not be able even to mention many collections, devoted to the work of individual scholars, the development of this or that discipline in a given country, that of the different fundamental concepts, like the number, function, infinitary magnitudes, differential, integral etc. etc., and the activities of the individual institutions, academies, societies, periodicals etc. At times, even a series of articles (for example, those of O.B. Sheinin on the history of the theory of probabilities and its applications) is of no less importance, than this or that book.

Let us conclude the retrospective of the historico-mathematical investigations here, and turn to a retrospective of mathematics itself. The problems selected herein for consideration, make no claim to completeness and, naturally, they express the interests of the author. As far as possible, the following exposition follows the chronological order of development of mathematics and, takes care of its regional specificities.

Neolithic Mathematics. Our judgements about the formation of the earliest mathematical notions, formed in the pre-historic times, are based on archeological data, sometimes, upon the written legends and inscriptions, preserved on the architectural structures and utensils, on the pictures found on the rock surfaces of the cave dwellings; and, finally, our ideas on the subject are also developed in analogy with the mathematical knowledge of those people and tribes, who are, or were a short while ago, situated at the lowest levels of cultural development. For some time now, the earliest developed culture known to us has been, the one that existed on the Indus river basin, in the middle of the 3rd millennium B.C.—the so-called Mohenjo-daro culture. Mathematical texts from this culture have not survived, and the inscriptions that have remained intact, have not been deciphered. We have extremely meager data about the arithmetical and geometrical knowledge of this culture, and still less—about its history; it appears to be close to the culture of Sumer. However, results of the investigations into the culture of Mohenjo-daro, are not in confirmity with the hypothesis regarding the introduction of this culture in the Indus Valley by some of the Aryan tribes (i.e. these studies do not confirm the hypothesis of B.L. van der Waerden; on this more later). Recently, an even more ancient culture has been discovered in Upper Egypt, but it remains almost uninvestigated.

The earliest preserved Egyptian representations of numbers date back to the first half of the 4th millennium B.C., but the two, aforementioned, so far preserved, basic mathematical papyri date back to the first centuries of the 2nd millennium B.C. The Babylonian cuneiform texts are divided into three basic categories: 1) the most ancient economic texts of Sumer, 2) the tables for multiplication, division and other operations, often also meteorological tables—dating back to the end of the 3rd millennium and, 3) some even later collections of problems—approximately belonging to the 9th-7th centuries B.C. All these written documents are close in time to the Indus Valley Civilization and, all of them go back to even earlier periods; whether or not there existed any direct contact between these civilizations, that however, remains to be established. A somewhat authentic information about the ancient Indian mathematics of the subsequent centuries, belong to a much later period; it is related to that epoch when the religious books—the Vedas—were composed. It is contained in some essays, enunciating the rules for the *construction* of sacrificial altars, in the so-called "Śulva-sūtras", written, probably, in the 6th and subsequent centuries B.C.; these have come down to us in several variants. The Chinese culture is also very ancient, but it is practically impossible to isolate the authentic facts from the legends, contained in the later Chinese chronicles. [For example, about the awareness of some particular instances of the theorem of Pythagoras in the 12th century B.C., and with its generalised form—in the 6th century B.C.] There is no doubt about the fact, that already in the school of Mozi, the philosopher and logician, i.e. in the 4th century B.C. or even earlier, the Chinese attained a high level of mathematical knowledge. By then, probably, many of those problems about which we came to know from the most ancient mathematical and mathematico-astronomical works—"Mathematics in Nine Books" and the "Treatise on Gnomon"—were already formulated and the methods of their solution were found; these books became famous through their editions published around the beginning of the Christian Era.

Sometime ago, in 1983, B.L. van der Waerden attempted a partial reconstruction of the mathematics of the Neolithic epoch. It predates the Egyptian, Babylonian, Indian and Chinese mathematics and serves as their most important primary source [24]. In the opinion of this leading algebraist and outstanding historian of science, there existed a very highly developed mathematics in the territories of Central Europe and Great Britain, somewhere between 3000 and 2500 B.C.; afterwards, this mathematics spread towards the South and the East, into the territories of Egypt, Babylon, India and China. Its traces are to be found in ancient Greece too—and this includes the works of Euclid and Diophantus; however, it was the Greeks who radically reorganised this ancient mathematics and created a deductive science based on definitions, postulates and axioms. It will not be possible for us, here, to enter into a detailed analysis of the ideas of van der Waerden. His book contains many interesting and valuable remarks, but his entire conception has been, on the whole, less convincingly argumented. Three basic arguments have been put forward in the author's introduction. The first among them—the presence of Pythagoras' theorem and of its application for transforming a rectangle into a square, in the "Śulva-sūlras", wherein the sides of the right-angled triangles used in the constructions are proportional to the "Pythagorean triplets" of natural numbers. "Pythagoras' theorem" and the extensive tables of Pythagorean number triplets were well known in ancient Babylon. From this, van der Waerden concludes—following A. Seidenberg (1978)—that there existed some kind of a common source, of the Babylonian algebra and geometry, the Greek geometrical algebra and, the Indian geometry. The second argument—the existence of a large number of similar problems in the ancient

Chinese "Nine Books of Mathematics" and in the ancient Babylonian texts, assuming in particular a knowledge of Pythagoras' theorem. The third argument—towards the end of the 70s, a number of archeologists studied some megalithic structures, erected on some platforms, for ceremonial rituals, as well as for definitely oriented astronomical observations. These platforms are bordered with menhirs, placed along circular, elliptical or oval lines or along the circumference of forms flattened out into circles. It is possible, some times, to determinately inscribe individual integral-numerical Pythagorean triangles into these figures. Such megalithic structures were erected since the middle of the 4th millennium B.C. and were widespread in Central Europe, Great Britain, Ireland etc., in the first half of the 3rd millennium B.C.; and, according to van der Waerden and Seidenberg, this testifies to the existence of a highly developed mathematics in the Neolithic epoch and it influenced the entire subsequent development of mathematics.

The decisive argument of van der Waerden is as follows: the discovery of the Pythagorean theorem and of the Phythagorean number triplets, were great discoveries, and the great discoveries of mathematics, physics and astronomy are, save in the rare cases, made only once; independent discovery of the said theorems and number triples in ancient Babylon (around 2000 B.C.), India, Greece (where they were well known not later than the 7th-6th centuries B.C.) and China, is improbable. Some unknown people took all these with them in the course of some migrations to the East.

This argument cements the entire conception of van der Waerden. It has been illustrated with the examples of momentarily invented epicycles and eccentric circles, of the establishment of the sphericity of Earth, the heliocentric system of Copernicus, the three laws of Kepler, and the laws of motion of Newton as well as his law of universal gravitation, the laws of optics etc. Independent discoveries—for example, of the non-Euclidean geometry by Lobachevsky, Gauss and Bolyai—are very rare. There is inexactitude in van der Waerden's enumeration; for example, R. Hooke discovered the law of universal gravitation independently of Newton; it is true, however, that he could not construct a system of celestial mechanics. The defect however, is not with the particular instances of inexactitude; independent discoveries are by no means a rarity in the history of mathematics and of the sciences in general. Here are some examples: the logarithmic tables of Napier and Briggs, the calculating machines of Schickard and Pascal, the analytical geometry of Descartes and Fermat, the differential and integral calculus of Newton and Leibniz, the theory of elliptical functions of Abel and Jacobi, Dedekind's and Zolotarev's theory of cut, the special theory of relativity of Einstein and Poincaré, Urysohn's and Menger's topological theory of measure... This list may be indefinitely extended further and, in general, in the given realm of questions, it is difficult to count and mutually compare the probabilities. One way or the other, according to van der Waerden, when a theorem like that of Pythagoras, is found in different countries, then the best course open is to accept the hypothesis of their dependence upon a primary source and to use it, as a heuristic principle.

It stands to reason, that the question of dependence or independence of identical discoveries in different cultural environs, requires to be investigated. Only this much is certain, that the solution of this question must not be based upon highly indeterminate probabilistic estimates and unprovable presuppositions about the course of development of humanity. Having put forward his hypothesis and heuristic principle, van der Waerden himself then and there notes many points of contact between the mathematics of China and Babylon or India and Greece; incidentally, these comparisons, made by him, are highly interesting and deserve serious attention. However, if such points of contact, yet to be studied in their full scope, did exist, then it is legitimate to ask oneself: were not the theorem of Pythagoras and the Pythagorean triplets born in the civilizations of Mesopotamia, from where they spread out in different directions? Why assume the existence of a highly developed Neolithic mathematics in Europe, in the 4th-3rd millennium B.C., about which we practically know nothing, when we know for certain that a Sumero-Babylonian mathematics did exist, which is known to us, at least in part? And what makes the hypothesis of a single source more preferable to the hypothesis of independent discovery of the theorem of Pythagoras, in course of the progress of architecture, that developed upon the ground reality of the general civic and ritual requirements of the people of a number of regions, which did attain similar levels of culture, at approximately the same time?

About the integral numerical Pythagorean triangles, which may be inscribed within the contours, along which menhirs were placed in a number of instances, it is not at all understandable, as to why the builders of the structures were in need of them. Traces of such triangles were not retained. Further, the contours themselves—be they spherical, flattened out and consisting of the arcs of circles of different radii, oval or even near elliptical—were outlined, one should think, with the help of simple siring contraptions. Right-angled triangles are not necessary for all such constructions.

Ancient Orient. While going over from the unwritten evidences of mathematics to the most ancient written mathematical documents, one must first of all turn to Egypt and Babylon. Here the information at our disposal is clearly fragmentary, but nevertheless, it does allow us to judge the systems of numeration, methods of computation and predominant types of problems of both the civilizations. It appears, that in Balylon, that component of mathematics was considerably more developed, which we can call algebraic: here we find cycles of problems expressed in quadratic equations or in systems reducible to them—the Egyptian papyri do not contain such problems; we have already mentioned the fact that the theorem of Pythagoras and the Pythagorean numerical triplets were known to the Balylonians. One of the Egyptian papyri dating back to the beginning of the 2nd millennium B.C., contains a simple algebraic problem, where the sum of the squares of the unknown quantities is given, there exits a given linear interconnection among the unknown quantities and the solutions—6, 8 and 10—happen to be double of the simplest Pythagorean numerical triplets; but the text does not mention any triangle, thus it would be hasty to conclude about any acquaintance with the general theorem of Pythagoras and with the Pythagorean numerical triplets in ancient Egypt.

During the last few decades, there were only a few substantial discoveries in this section of the history of mathematics: the most interesting among them being the "Balylonian numbers" detected by A.A. Vaiman (1957)—these are numerical triplets, expressing the ratios of length of some line segments, parallel to the bases of a trapezium, which divide it into paired bands of equal area; these numbers turn out to be Pythagorean numbers, their sums, and differences. But the study of the mathematics of Babylon and Egypt produced a number of interesting reconstructions of those methods with the help of which various problems were solved there. In both the cases, the texts contain only calculations, providing the solution or even, straight off, the answer, without any explanation: the enunciation is prescriptive in character and does not include any such element, which we would have now termed theoretic. It is clear, however, that the solutions of many problems could not have been obtained purely empirically. This is true, for example, about the rule for calculating the volume of a truncated pyramid (Egypt) or, about the solution of quadratic equations (Babylon); it is not likely, that the rule for the summation of the sequence of the squares of natural numbers and, the correlation among the Babylonian and Pythagorean numerical triplets (Babylon) or, the (approximate) equality of the area of a circle with that of a square, a side of which is equal to 8/9 of the diameter of that circle, were detected accidentally. All the texts of the ancient Oriental mathematics, known to us, highlight only one side of it: these are either manuals for solving a definite type of problem, or collections of exercises with answers, and sometimes with verifications. There is no doubt, that there were mathematicians with a command over the tools of arithmetical, algebraic or geometrical deductions. This apart, we do know that when the texts under consideration were being composed, mathematics had already attained such a level of development, that apart from the problems generated by the direct requirements of economic, political and technological practice, it also handled those problems which were bereft of all practical significance, those which arose in course of the development of mathematics itself. The computation of the area of a rectangle with given sides, is an elementary practical problem, solved arithmetically. Reflection on it gives rise to an abstract algebraic problem: the determination of the sides of a rectangle in terms of a given area and perimeter.

Some historians of science (A. Szabó, I.G. Bashmakova, P.P. Gaidenko and others) put forward the view, that only in Ancient Greece did mathematics arise as a science, when significant portions of this discipline began to be constructed in the form of deductive axiomatic systems: They opine, that there exists no ground for thinking, that mathematical knowledge was formulated in ancient Egypt or Babylon, into such systems; there, many results were obtained empirically or through unsubstantiated generalization from particular modes of calculation or measurement. Notwithstanding this, the following remark of N. Bourbaki [21] is fully convincing; it is not possible to view the entirety of Babylonian algebra as a simple collection of exercises, solved empirically, to the touch, and if it does not contain any "proof" in the formal sense of the word, then some sort of, not yet fully realized, logical arguments were put forward. By using the analogue of a mathematical terminology, used in another context. We may call ancient Oriental mathematics—"piece-wise deductive".

It has already been said in the beginning of this paper, that the words "proof" and "science" do not have any univocal meaning when they are viewed in their history, different contents were put into them at different times. The refusal to call the mathematics of Egypt and Babylon—science, because there are no proofs in their written documents, is as groundless, as the exclusion of impressionism or the abstract school from painting, because they are "not realistic", and of the works of Éluard or Khlebnikov from poetry, as these do not resemble those of Verlaine or Blok. The same is true also in respect of the ancient mathematical texts of China and India.

Ancient Greece: Hellenism. After Egypt and Babylon, it is natural to turn to Ancient Greece. There are several aspects of the problem of the sources of Greek mathematics, the first among them being the question of Oriental influences. Taking it up in the case of Egypt, B.L. van der Waerden now, as they say, elevates it, in the ultimate analysis, to a hypothetical European culture of the Neolithic epoch; O. Neugebauer takes into consideration the emergence of the theories of irrationality, proportions and integrations within Greece and thinks that the Greek geometrical algebra shows Babylonian influence, which became stronger in the beginning of Hellenism, and he raises doubts about the roles of the Ionic school and of Pythagoras; I.G. Bashmakova is of the opinion that Pythagoras is the creator of mathematics as a science; A. Szabó gives precedence to the influence of the Eleatic school and the introduction of the rule of contraries; Leonid Yakovlevich Zhmud has recently raised doubts about the presence of Oriental influence ... It is apparent that opinions are changing and, clearly, the discussion around this question will continue. Perhaps, the question of formation of Greek mathematics should be considered within the wider frame-work of social and ideological development of the entire Mediterranean culture. Here the reader is recommended to get acquainted with the materials published in the collection : "Metodologicheskie problemy razvitiya i primeneniya matematiki" ["Methodological Problems of Development and Application of Mathematics"] (M., 1985), especially with the section: "Metodologicheskie aspekty stanovleniya matematicheskovo znaniya" ["Methodological Aspects of Formation of Mathematical Knowledge"]

While dealing with the problem of formation of the mathematical deductive method, in that specific form, which it assumed in Ancient Greece, i.e. in the first place, in the axiomatization of geometry (but not of arithmetic, the reason behind which has been investigated by S.A. Yanovskaya in 1958), from the very first steps one runs against the non-univocality of the possible interpretations of the ancient idea of the infinite, especially in the early stages, and of the corresponding terminology. What gave rise to this idea? How to understand the "apeiron" of Anaximander [apeiron—a concept introduced by Anaximander of Miletus (c. 610-546 B.C.) to denote boundless, indefinite, qualityless matter in a state of constant motion—Tr.] or the "aporias" of Zeno of Elea (c. 490-430 B.C.) [aporia—a problem which is difficult to solve, owing to some contradiction in the object itself or in the concept of it—Tr.]? There is no doubt about the need for discussing these questions, but we are still far away from their unanimous solution.

Having mentioned the names of Anaximander and Zeno of Elea, one has to state that the problem of infinite did have a decisive impact upon the entire methodology of Greek mathematics, upon its various aspects. 60 years ago H. Weyl wrote, that mathematics is the science of infinity and if an intuition of the infinite was characteristic of the Oriental world, where it did not give rise to any question, then the Greeks reconstructed the polar opposition of the finites and the infinites into powerful instruments for the cognition of reality. Unfortunately, the opinions about the problem of infinity and about the infinitesimal methods of the ancient Greeks, are so divergent, that one has to refrain from characterizing them in the present paper.

In the recent times, unfortunately, the study, almost only, of the "Arithmetic" of Diophantus, has come to occupy one of the foremost positions in the study of the history of Hellenic mathematics. I.G. Bashmakova was the first to produce a deeper study of this work, utilising the tools of modern algebraic geometry (1972 and subsequent years); soon it was independently extended to the study of R. Rashed—on the so called Arab Diophantus (1974 and subsequent years), and independently by J. Sesiano and, to that of the so-called "Diophantine analysis" upto the epoch of Fermat (I.G. Bashmakova and E.I. Slavutin, 1984); these studies threw a new light upon the formation of this field of mathematics, which played an important role in the development of theory of numbers and algebra. In a recent book on the development of the theory of numbers (1983), A. Weil has related the entire work on indeterminate analysis upto the time of Viète [1540-1603] and Bachet [1581-1638], with the pre-history of the theory of numbers, since, therein, the attention was fixed, not only on the search for the integral, but also on that for the rational solutions. Weil has considered the works of Diophantus mainly in connection with those of Fermat, whom he considers to be one of the founders of the modern theory of numbers, together with Euler, Lagrange and Legendre.

The "Arithmetic" of Diophantus exerted a direct or (and) indirect influence—mediated through some unknown links—upon the development of Arab algebra. It has been a great influence. We must mention here the fact, that so far the very emergence of the "Arithmetic", has almost always been viewed as an isolated event in the development of the mathematics of the Alexandrine epoch. This work determined a trend of thought different from the classical one, proposed by Euclid, Archimedes and Apollonius. To all appearance, it was the result of a synthesis of the Classical and the Oriental traditions; the creation of the empire of Alexander of Macedonia and, after its fall—the emergence of several Hellenic states, created the general historical preconditions for this synthesis. There are common elements in the works of Diophantus and in those of Hero of Alexandria, who lived nearly two centuries before the former. However, there still exists a very big gap in the primary sources known to us, which cannot be filled by the data on hand about the problem of piers, related with the name of Archimedes. The ancient method of solving this problem is still not known to us. The possible connection of "Diophantine Analysis" with the mathematics of India and China also remains unknown. In India and China too, the integral and rational numerical solutions of different kinds of indeterminate algebraic equations, did occupy an important place.

The Middle Ages. As we go over to the Middle Ages, we must first of all state that this term is unsatisfactory, even if because of the fact that in different regions its natural boundaries belong to different centuries. In the absence of a better name, however, we shall be using this term. In places it began at the junction of the pre-christian and the christian eras, at others even later and, came to an end in the 15th- 16th centuries. The states that existed in this period, over considerable parts of the territories of China, India, the Arab countries and Europe, were in the main similar types of feudal (?—Tr,) economic and political formations [while discussing the ancient and the middle ages, the question of the Asiatic etc. modes of production cannot be/should not be avoided—Tr.]. They attained almost the same levels of technical and material culture. It stands to reason, that the exchanges of material and spiritual values that took place among these regions, were by no means regular and, were interrupted by wars and internal disorders. A natural consequence of all this has been the emergence of similar practical problems before the mathematics of these four regions. In some of the works—written in the period 1958-1961 by the present author, often in collaboration with B.A. Rozenfeld—a conception of medieval mathematics has been proposed, wherein it has been considered as a single whole; but the specificities inherent to the mathematics of each of these regions have been taken note of there; these specifics were largely the consequences of even earlier scientific, philosophical and religious traditions prevalent in these regions. That is why, it would be better to consider the changes that have taken place in the retrospective of mathematics, during the last few decades, separately, within the frame-work of each region. It must be mentioned initially, that during the last few decades considerable success has been achieved in the study of the medieval mathematics of many Oriental countries; and, to a great extent this was made possible by the decolonization of the territories under the control of the imperial powers, the emergence of independent states in Asia and Africa, as well as owing to the fast progress in those Republics of Central Asia and of the Caucasus, that were backward areas before the October revolution in Russia and, were in the best of circumstances—second grade areas of the Russian empire.

China. Our knowledge of the development of mathematics in China has greatly increased in the recent years. European scholars obtained their first solid informations about Chinese mathematics through an English language book (1913) by the Japanese scholar I. Mikami. In the first half of the 20th Century, work was conducted in this field in Europe and in China, mainly independently of each other. During the 30s-60s, considerable contributions were made by Li Yan and Qian Baocong, and by Mikami, who continued his investigations; however, they wrote mainly in Chinese and Japanese and, for a long time their books and papers were accessible to only a few European or American historians of science (now the number of sinologists and historians of mathematics have increased). That is why the publication of a generalised work on the mathematics of China, by the English sinologist (and biologist) J. Needham, was an event of great significance; Needham wrote this book (1959) in collaboration with the Chinese mathematician Wang Ling. Later on the work gathered momentum in China and in Europe. Many of the classical works of Chinese mathematics were translated in the European languages, and books were written about the great mathematicians of China. From among these translations one must first of all, mention the Russian edition of the tracts of the so-called "Ten Books", which have been provided with commentaries. The last of these "Ten Books" was written in the 7th century and, the earlier "Nine Books of Mathematics" were published some where near the beginning of the Christian Era. E.I. Bereozkina's translations of these works were published in the period 1957-1985; and in 1980 she initially summed up the results of her investigations in a special monograph, wherein she has considered the attainments of the mathematicians of China upto the beginning of the 14th century, but somewhat more briefly. K. Vogel prepared a German translation of the "Nine Books of Mathematics" (1968) and, this is accompanied by his own commentaries. A collection of papers on the same book and on its most important commentary, composed in the 3rd century by Liu Hui, has been published in 1982, in the Chinese language, along with an English resume. The authors of these papers—Bai Shangshu, Li Di, Shen Kangshen and others—made a significant contribution to a fuller study of the "Nine Books of Mathematics". [The present author is greatful to A.K. Volkov, for translating parts of the big chapters of this collection, as well as Shen Kangshen's review (1985) of E.I. Bereozkina's book (1980) mentioned above—from Chinese. Kangshen has justly stressed the importance of the commentaries of Liu Hui, in his review of Bereozkina.] For nearly one and a half thousand years the mathematicians of China, on very many instances, took their cue from these "Nine Books of Mathematics", and this explains the special attention that has been paid to it; see: the bibliography prepared by the German sinologist Hermann Kogelschatz (1981).

The epoch of pre-decline flowering of mathematics—above all of algebra and theory of numbers—in ancient China of 13th century, has been the subject matter of a number of important investigations. A Belgian scholar U. Libbrecht (1973) made a detailed study of a treatise by Qin Jiushao; Lam Lay Yong, who works in Singapore, published (1977) an English translation of a treatise by Yang Hui; J. Hoe (1977) published a French translation of the works of Zhu Shijie, and used therein the special symbolism devised in that epoch. K. Chemla (1982), just like Hoe, used a semi-symbolic language, in her doctoral dissertation on an algebraic treatise of Li Ye—a contemporary of Qin Jiushao. Unfortunately, this dissertation, as well as a much earlier work by another French sinologist R. Schrimpf (1963), on the mathematics in China upto the 7th century, has not been published.

In this connection, special mention must be made of the papers of Ho Peng Yoke (Malaya) on Qin Jiushao, Zhu Shijie, Li Zhi (or Li Ye), Liu Hui and Yang Hui, in the 3rd, 8th and 14th volumes of the American Dictionary of Scientific Biography [22].

As a result of all these investigations our knowledge of the development of mathematics in China, has been greatly extended. It appears now that its arithmetico-algebraic component is richer than what it was thought to be, in the beginning of this century. At the same time, a number of questions remain unsolved and far from all the important primary sources have been studied till date. Undoubtedly, there had been mutual interaction among the plural mathematics of China, India and the Arab countries, which was reflected even in Europe: the migration of similar or identical problems, and not merely the deductions constructed upon the shaky principle of *post hoc ergo propter hoc*, testify to this effect. Modem historians of mathematics do take note of all these moments, but by far not all of them are clear and definitive.

C.P. Luckey has advanced a hypothesis about the transference of the rule of two false positions and of the methods of extracting the square roots and cube roots of rational numbers, to the Arab countries from China; but it lacks certitude. Perhaps the Belgian sinologist L. van Heé was the first to put forward the idea, prevalent till recent times, that Liu Hui solved the problem of measuring inaccessible objects and the distances upto them, by basing himself upon the similarity of triangles. According to a recently advanced hypothesis, Liu Hui used some methods which are characteristic of the Greek geometrical algebra. We have already mentioned the noteworthy proximity, of the treatment of the basic ideas of geometry, in one of the works of the Moist school 4th century B.C., with a somewhat earlier ancient Greek treatment of the same. Exactly in the same way, the similarity between Liu Hui's methods of approximate calculation and those found in Archimedes' "Measurement of Circle"—notwithstanding their inessential technical differences—has been mentioned more than once. All of this leads to an idea about the existence of scientific contacts between China and the Hellenic countries: trade was conducted between China and the Roman empire. In the more recent years an observation of D.B. Wagner (1978), to the effect that the ancient Chinese mathematicians used the so-called principle of Cavalieri, in their studies of the problem of cubature of a sphere, has become an object of special interest in Europe; apparently, it was known much earlier, to Mikami (I have not yet been able to verify it). The history of this question is as follows. Liu Hui expressed the volume of a sphere in terms of the volume of a body, contained within the surfaces of two cylinders, inscribed in a cube and having mutually perpendicular axes, but he could not determine the volume of this body. It is very likely that Liu Hui used the so-called principle of Cavalieri, though this principle has not been formulated in any of his texts known to us. We first come across a formulation of this principle in the 5th century, in the writings of Zu Chongzhi. [It is difficult to translate this formulation with exactitude, as the corresponding terms do not have fully determinate mathematical significance in latter Chinese speech; while referring to Anaximander's concept of "apeiron", we have already mentioned the possibility of non-univocal understanding and translation of the ancient terms.] Zu Chongzhi's son Zu Geng applied this principle and found out that the volume of such a body is equal to the 2/3 of the cube, which gives us the cubature of a sphere. It is remarkable that the result of Zu Geng has been formulated in the 2nd proposition of Archimedes' "Epistle to Eratosthenes"; in this work the method of indivisibles has been regularly used for heuristic purposes (but not for obtaining any "strict" proof); unfortunately, the conclusion of this proposition, contained at the end of the "Epistle", is not known to us. We do not find a statement of the "Principle of Cavalieri" in the Greek texts known to us, however, Archimedes' quadrature of the ellipse, viewed as the result of compressing a circle, leads to the thought, that in essence, this principle had been used by him intuitively. Personally to me it seems plausible, that there had been a Greek influence upon the infinitesimal methods of the Chinese mathematicians of the 3rd-5th centuries.

On this score Lam and Shen Kangshen display greater restraint (1985), when they say that Liu Hui's text contains no evidence of a Greek influence; such an influence is discernible only later on, in the works of Mei Wending—a mathematician active in the 17th-18th centuries.

On the whole, however significant may the achievements in the study of Chinese primary sources be, still a very great amount of work remains ahead of us. Perhaps one of our primary tasks is to study all the hitherto known commentaries on the "Nine Books of Mathematics", which shed light upon the methods employed for solving those various problems, which we now classify as algebraic, number-theoretical and geometrical. The Chinese treatises, beginning with the "Nine Books of Mathematics", do not contain any proof, there we find only laconic mention of the methods of solving the problems; the substantiations of these methods are often met with in the commentaries. More than thirty years ago I expressed the opinion, that it would be unjust to judge the mathematics of China on the basis of its collections of exercises, that here and in the case of the mathematics of ancient Orient (and, I add, of India), we must distinguish between the manner of presenting a discipline in the textbooks, from the creative elaboration of the methods of investigation which precedes it. Both in China and in Europe, the study of these commentaries are, in essence, in their early stage. Recently, A.K. Volkov examined the commentaries on the rules for calculating some areas (1985) and, therein he noted that in the mathematics of ancient China, the very concept of "proof" and the "systems of proofs", have their own specific characteristics: not an axiomatic theory, but a theory of models happens to be a more adequate analogue of the ancient Chinese logical system, and their criteria for deciding about the correctness of propositions correspond to this; this question deserves a more detailed study. The intensive work that is being carried out in this field in China, USSR (now erstwhile—Tr.), France, FRG (now Germany—Tr.), and in the other countries, will soon yield new results and, one may say, newer postulation of the problems too. The development of mathematics in China beyond the classical period, i.e. in the 14th century and afterwards, has not at all been touched upon here.

India. While dealing with B.L. van der Waerden's hypothesis about Neolithic mathematics, we have already mentioned the latest investigations on the history of mathematics in India. Apart from the more detailed analysis of the works of individual mathematicians like Śrīdhara or Mahāvīra (A.I. Volodarsky, 1966, 1969), the reconstruction of the solutions of some problems in the Āpastamba

"Śuba-sūtras" (A.E. Raik and V.N. Ilyin, 1974), the recent observations of P. Singh (1985) about the so-called Fibonacci numbers of 7th-8th century Indian mathematics or, the works on the history of Indian astronomy by D. Pingree (USA, 1963) and A.K. Bag (India, 1966-), the most interesting attempts in this field were concerned with the exact determination of the connections of Indian science with the science of the other regions and, with its place in the overall progress of mathematics. The aforementioned book by A.I. Volodarsky contains an overall survey of the work done prior to 1977. [Here the reference is to: *Volodarsky A.I.* Ocherki srednevekovoi indiiskoi matematiki (Esays on Medieval Indian Mathematics). M.: Nauka, 1977, 182 pages.—Tr.] It appears that, comparative analysis must be further continued in this direction. It is a fact, that here the investigator has to face a deficiency of exact information, so much so, that even the emergence and the earlier stages of development of the now commonly accepted system of decimal positional numeration, remain largely unclear.

Arab Mathematics. During the last quarter century, our ideas about medieval Arab mathematics have changed no less substantially, than those about mathematics in China. The term Arab Mathematics has struck roots and comes in handy, though in fact, at issue here is the mathematics of the people of many countries [where at a given period of time Arabic has been the principal language of science—Tr.], stretching from the Pyrenean peninsula, through the northern, Mediterranean regions of Africa, Near East, Central Asia and further ahead, roughly upto the present borders of China and India [in the case of India it should mean upto the India-Burma border of Mughal India and, in case of China—at least upto the eastern borders of Xinjiang or Chinese Turkestan—Tr.]. The sources of Arab mathematics remain largely uninvestigated, they go back to the mathematics of Babylon and Egypt, a large number of Hellenic states, Byzantine, and to the science of ancient Khwarezm, not to speak of the much later influences which came through the contacts with China, India etc. Here one may put forward a number of questions, but in the absence of exact data, it would be more than difficult to answer them. We have left out such questions in the present paper. However, one must at once make one observation about the expression "Arab Mathematics". We are unable to find a similarly brief expression, which may be a substitute for it; but one has to stress the fact that even a brief bibliographical survey of the original literature shows the special significance of the contributions of the mathematicians of Central Asia and, this justifies a separate treatment of the history of mathematics of Central Asia of the period under consideration, in a number of books. When one deals with the cultural developments of the Central Asian Republics of the USSR [now CIS—Tr.], which were, quite understandably, closely connected with the culture of those legions, where Arabic or Persian had been the principal language of the scholars, then such a separate treatment becomes essential.

A more detailed study of the already known Arab works, and, to a greater extent, an analysis of a large number of mathematical manuscripts preserved in the various libraries and archives, showed that medieval Arab mathematics did attain a scientific level, which is much higher than what it was earlier thought to be. It goes without saying, that the mastery of the Greek scientific heritage, which was one of the consequences of Arab expansion and of the formation of the Arab states in the territories that were earlier under the rule of Rome, was of great significance for the progress of scientific and philosophical ideas in these states, which was often supported (and sometimes opposed) by the rulers of these states as they changed hands. It is enough to mention the scientific school of Bagdad, of the end of the 8th-9th centuries, which blossomed soon after the stabilization of the Caliphate of Bagdad and, the Samarkhand school of the first half of 15th century, during the rule of Ulugh Beg. Thanks to the opportunity of quick assimilation of the heritage of Greek ideas and the very turn of thinking, the science of the Caliphate of Bagdad found itself in a situation that was much more favourable than the one in which science found itself in India and, even more compared to the situation in far off China. However, an entirely wrong approach to Arab science, including Arab mathematics is prevalent till date; according to this interpretation Arab science is nothing but a transmission point between Greece and Rome on one side and the Europe of the middle ages and of the beginning of the modern times on the other. This conception was clearly formulated by E. Renan, more than hundred years ago, in 1863. It was he who put the expression "the Greek miracle" into circulation and, considered Arab science to be a reflection of Greek science, combined with the influences which came from Persia and India. In defence of his thesis Renan adduced even philological considerations: he, and not he alone, thought that the Indo-European languages are more suitable for expressing abstract concepts, than the Semitic ones etc. However, Renan was not an historian of science; but his ideas were developed by such important specialists in the field as P. Tannery and P. Duhem; these opinions are shared by some of the leading scholars even to-day. This Eurocentric conception of history of science, does not correspond to the actual course of scientific progress, and it has been criticised many a times in the Soviet, as well as in foreign literature. What is special about Arab mathematics is this, that here we find a magnificent development of certain trends which originated in Greek or Indian mathematics and, some important advances in new directions, It would be enough to cite some examples:

The first systematic construction of the decimal positional arithmetic, the principles of which were, however, borrowed from the Hindus. Introduction of decimal fractions and the method of extracting the n-th roots, by using binomial expansions (11th-15th centuries); the different numerical methods for solving algebraic equations, and besides this, an example of approximate solution of a transcendental equation, with the help of successive iterations, extension of Diophantme analysis, solution of indeterminate linear Systems, properties of the amicable numbers, Wilson's theorem (9th-10th centuries).

An original theory of ratios and proportions; extension of the concept of number to the positive irrational numbers; arithmetization of the ancient teachings on quadratic and biquadratic irrationalities (9th-13th centuries).

Separation of numerical algebra together with the algebra of polynomials as an independent discipline; a developed geometrical theory of cubic equations and, a geometrical theory of the equations of fourth power (15th century, the corresponding treatise is yet to be traced).

Different theories of the parallels, connected with the attempts to prove the 5th postulate of Euclid (9th-13th centuries).

A reconstruction of the 8th book of the "Conic Sections" of Apollonius (9th century).

New quadratures and cubatures (9th-10th centuries).

In this list we have not specially isolated those trends and results, where the Arab mathematicians happened to be pioneers; it is enough to state that where they broke entirely new trails, they went considerably further than their predecessors. We have neither mentioned the names of those mathematicians, nor the names of those historians of science, who have of late elaborated or are continuing to elaborate upon the entirety of this vast complex of disciplines, theories and problems: the lists of either of them would be very large; one may find these names in the corresponding literature. [See, for example: *Matvievskaya G.P.*, *Rozenfeld B.A.*, Matematiki i astronomy musuilmanskovo srednevekoviya i ikh trudy (VIII-XVII vv.)/Mathematicians and Astronomers of the Muslim Middle Ages and their works (8th-17th centuries)/; in 3 volumes (479+650+372 pages), M.: Nauka, 1983.—Tr.]. However, in view of the special importance of the question of evaluation of the role of Arab mathematics in the subsequent forward movement of mathematics—a question, which has already been touched upon—it is essential to dwell upon it. As we have noted above, Renan's evaluation of the issue, continues to find its supporters even in our time. B.L. van der Waerden sticks to a clear cut Eurocentric position. He came forward with the following sketch of the emergence of modern science, in a seminar held in Oxford in 1961 [25], while discussing J. Needham's paper on science in China. He said, Newton's mechanics is the basis of modern science, wherein three threads remain intertwined—each of which has emanated from Greece. The first among them is the planetary astronomy, leading to Copernicus and Kepler, i.e. to the necessary prerequisites of Newton's mechanics. The second thread: Newton's use of Apollonius' conic sections and of the entire structure of the Greek axiomatic geometry, which served as the model for Newton's axiomatic mechanics. The third thread emerges from the Greek mechanics of Archimedes and of some other scholars. There exists no doubt about the enormous significance of the Greek heritage for the sciences of modern Europe, and that includes the work of Newton. However, only the Greek heritage was not enough for it. The discoveries of Copernicus, Kepler and Newton are essentially based upon the Arab traditions of a highly developed new algebra and trigonometric system. The intertwining of the Greek and the Arab threads of scientific development and their subsequent creative synthesis in the mathematics, mechanics and astronomy of medieval Europe, had provided the necessary prerequisites here. Not all the attainments of the sciences of the Arab countries, were known in medieval Europe, and a lot of it had to be discovered anew. However, often, even a fragmentary introduction of the results of the Arab investigations, served as points of departure for important trends in modern European mathematical thought. Such, for example, has been the case with the Arab theory of parallel lines, an acquaintance with which in the 17th century played an important role in the first stage of evolution of the non-Euclidean geometry.

All the same, if the sources of the sciences of modern Europe go back not only to Ancient Greece, but also to the countries of the East, and the terms "western science" and "eastern science" become admissible only with some stipulations—the mathematics of medieval Europe did have its own specificities, which were only very insignificantly, or not at all, characteristic of the other times or regions. Two of these had a very important or even determining significance for the formation of the mathematics of the modern period.

Here, first of all, we have in view, the creation and systematic perfecting of symbolic algebra, in the 13th-16th centuries. The timid steps taken in this direction, in the Moorish countries, are not going to be taken into account here: thanks to the march of world history, these steps could not be continued, and they failed to exert any influence upon the subsequent progress of mathematics. The formation of symbolic algebra was of immense significance for the entire further development of mathematics, and for developments beyond its boundaries; it was Leibniz who first evaluated the role of symbolism in human thought. Idea- and time-wise, the progress of symbolic algebra came along with such attainments of the 16th century, as the solution of the equations of 3rd and 4th power into radicals and, the introduction of imaginary numbers. Here the mathematics of Europe broke an unbeaten trail and this led to results of truly universal significance for the entire system of physico-mathematical sciences.

Another characteristic specificity of the mathematics of medieval Europe was connected with the distinctive development of some ancient natural-philosophic and scientific ideas, which, to a significant extent, go back, on the one hand to Aristotle and his school, and on the other—to Pythagoreanism and to Plato. Here, medieval (European) mathematical thought went far beyond the boundaries of that elementary mathematics, which was then known in all the four regions considered in this paper. On the one hand, it was the programme of mathematization of the entire world of knowledge, put forward by the scholar from Oxford R. Grosseteste and his pupil R. Bacon, and together with it the development of experimental method and of the technical means of scientific investigations. On the other hand, it was the first, but already fully perceptible growth of the infinitesimal mathematics of a new type, elaborated first of all in the universities of Oxford and Paris (Sorbonne). An in principle new development of the infinitesimal ideas took place here, and then also in the other universities of Europe. Together with this renewal and the deepening of the ancient discussions about the nature of the infinite in both of its forms, of the potential and the actual infinity, continuity and discreteness etc., within a short while, as has been pointed out by N. Bourbaki, the foundations of a theory of change of magnitudes, viewed as functions of time, and of their graphic representation, were laid down—true, in a rudimentary form. The English (T. Bradwardine, R. Swineshead etc.) and the French (especially, N. Oresme) scholars of the 14th century made a bold attempt to quantify the basically qualitative natural philosophy of the Peripatetics, with the help of infinitesimal ideas. First of all, a new interpretation was given to those sections of Aristotle's physics, wherein the interrelations of force and motion and, force and resistance has been considered—and this turned out to be especially important for further developments; in other words, a reconstruction of the Peripatetic mechanics was undertaken; after that, all kind of changes of the continuous, and partly also of the piece-wise broken, measurable quantities or, in the terminology of the Peripatetics, the intensification—strengthening and remission—weakening of all kinds of "forms" or qualities, like heat, colour etc., as well as of goodness, sin etc.—the varying intensity of which were dependent upon their extensity—the spread of their intensity over finite or infinite intervals in space or time, were subjected to mathematical treatment. Here the simplest of mechanical movement, i.e. spatial displacement too belongs to the category of form. Generally speaking, not theological or ethical, but rather natural-scientific, and in the first place mechanical intensity, is at the centre of interests here.

Quite a vast literature has been devoted to these theories of the Oxford and Paris schools, which were extremely close idea-wise, though coloured, so to say, in different tones (the English worked out their "calculations" at a more abstract-quantitative level; and the French "theory of widths and lengths of forms" made wide use of the graphic representations, which, however, were not alien to the "calculators"). The works of P. Duhem, published in the beginning of the 20th century, laid the foundations of the studies in this field; however, V.P, Zubov (1948) has convincingly demonstrated that Duhem was not at all impartial in his judgements. Here, once again, it is not possible to go into the details, and, by way of evaluating the teachings under consideration, it is enough to state that, therein we already find the formation of an idea of variability—flow (fluxus) of magnitudes, of momentary speed and acceleration, for which suitable, even Latin, terms were introduced and, the basic law and the other properties of uniformly accelerating motion were proved at an entirely abstract level, not connected with physics.

The calculations and the theory of widths and lengths of forms became quite widely known in the 15th and 16th centuries, first through manuscripts, and then through printed publications and, in the university-level teaching of a number of countries. During the last quarter of a century, a very large number of investigations have been devoted to this trend of medieval European mathematics, and this includes the publication of many manuscripts; together with this, the influence exerted by this trend upon the formation of the "mathematics of variable quantities" (A.N. Kolmogorov's term) in the 17th century—on Galileo, Napier, Barrow, Newton and his school, and very likely also on Descartes and Leibniz—has been studied in greater detail. Of course, there is no question about the beginnings of analytical geometry and of infinitesimal analysis, of laying their foundations, in the schools of Oxford and Paris; at issue here are the anticipations of and the ideational preparations for the just mentioned sections of mathematics, the foundations of which were laid in the 17th century. In particular, such ideational connections are evident in the terminology of Newton's "method of fluxions" and, in the fact that till date, in different languages, we use the expression "flowing coordinates" (the terms "variable", "function" and "coordinates" were introduced by Leibniz). Till date, the works of V.P. Zubov (1962 and 1965) remain the best Russian work in this field (the second book was published posthumously) but, naturally, the results of the later investigations could not be included in them.

Modern Age. In what follows, we shall be providing an even more fragmentary survey of the changes, that our notions about the historical past of mathematics have undergone, and we shall be illustrating it with only a few examples. Choice of the examples will be, to a considerable extent, connected with the recent archival investigations and, they are aimed at showing how these investigations are important for making our knowledge, not only of medieval mathematics, but also of the mathematics of the modern and recent times, more exact.

One of the greatest events in this area has been the publication of the eight volumes of the mathematical manuscripts of Newton, edited by D.T. Whiteside and his colleagues (1967-1981). This edition contains excellent commentaries. It has fundamentally changed our ideas about the scientific career of Newton and about the chronology of his discoveries. We have also come to know of those of his discoveries, which remained unpublished in his lifetime, owing to various, and not always clear, reasons. Thus, Newton discovered the expansions of Taylor and MacLaurin; he was the first to attempt an axiomatization of the method of fluxions; he proposed remarkable examples of asymptotic serial expansions etc. Almost simultaneously, A.R. Hall brought out a 7-volume edition of the complete correspondence of Newton (1959-1977). Here. V. Boss' book on the spread of the ideas and discoveries of Newton in 18th century Russia (1972), deserves special mention. <For the current state of Newton studies go to: http://www.newtonproject.ox.ac.uk/>

The work on the scientific legacy of Leibniz has been less successful; often his manuscripts are found to be chaotic in character and can be read only with great difficulty. [There are nearly 75000 separate works of Leibniz, preserved in the Leibniz Archives of Hanover, and many of them remain unpublished till date. On this see: Katolin L., "Mee byli togda derzhkimi parnyami M., "Znanie", 1979, p. 70.—Tr.] Study of Leibniz's legacy began long ago, but, in spite of many interesting results obtained so far, the principal work remains ahead of us. Some of the important relevant publications in the field are: the 1st volume of Leibniz's mathematical, natural scientific and technical correspondence, pertaining to the period 1672-1676, published by J.E. Hofmann (1976); the same Hofmann prepared a detailed name index to the entirety of Leibniz's correspondence (1977); E. Knobloch published a dialogue by Leibniz, which contained, among other things, the first clear expression of the idea of multidimensional space (1976); three volumes of the publications and investigations of E. Knobloch (1973-1980) are devoted to Leibniz's works on combinatorics and theory of determinants; the two volumes of the papers read at the Leibniz-Seminars of Hanover, 1966 (1969) and Paris (1978); and the complete chronicle of the life and work of Leibniz, prepared by K. Müller and G. Krönert (1969). Some of the published works in the field deal with Leibniz's treatment of the problem of foundations of the differential calculus; here, the emergence and development of non-standard analysis has led to a re-assessment of the earlier evaluations of the relevant contributions of Leibniz; this holds good for Newton too. Finally, one must mention A.B. Shtykan's (1952) investigations on the first analog integrators, invented by Leibniz and Newton, as well as by Ch. Huygens and Joh. Bernoulli. Information about the current status of publication of the works of Leibniz is available at: <https://leibnizedition.de/de/>.

Here one must also mention the deep-going investigations on the history of Jacob Bernoulli's work on the theory of probabilities—based, to a considerable extent upon archival materials—and the publication of his fundamental work in this field of mathematics the *Ars Conjectandi* [*The Art of Conjecturing*] (as well as those related to the early stage of the spread of his ideas), conducted by B.L. van der Waerden, K. Kohli and J. Henny (1975). J.V. Uspensky's Russian translation of the basic theoretical part of this work, which was subjected to deep going analysis by A.A. Markov (1913), has been reissued recently with Yu. V. Prokhorov's and O.B. Sheinin's commentaries (1986); A. Hald (1984) and the present author (1986) too dealt with a number of related questions.

Limiting ourselves only to the archival legacy of the most outstanding mathematicians, we must now go over to L. Euler. Here a great amount of work has been done involving close cooperation among the scholars of USSR, GDR, France and Switzerland. In view of the variety of the work done, we shall have to limit ourselves to listing the basic publications. These are: a complete description of the materials pertaining to Euler preserved in the archives of the Academy of Sciences of the USSR (1962) and a part of his scientific diaries, a complete description of the materials preserved in the archives of the Academy of Sciences of the GDR (1984), 3 volumes of Euler's letters to the Peterburg Academy of Sciences from Berlin (1959-1976), an annotated index of the complete correspondence of Euler—published in Russian (1967) and, a considerably supplemented German edition of it—published as volume I of series IV of the Complete Collected Works of Euler (1975). Volume V of this series contains the correspondence with Clairaut, d'Alembert and Lagrange (1980), and Volume VI—the correspondence with Maupertuis and Freidrich II (1986). So far, volume IV has been issued in another edition, which contains the correspondence with Goldbach (1976). All these volumes are being supplied with a large apparatus of commentaries; the publication of series IV continues. These fuller studies of the archival materials pertaining to Euler and their publication has been connected with some memorial years related to his life—1957, 1982 and 1983 [L. Euler was born in 1707 and he died in 1783, thus his 250th birth anniversary fell on 1957, 275th birth anniversary—on 1982 and, 200th anniversary of his death—on 1983.—Tr.] and with the holding of various conferences and meetings, as well as with the publication of some collections containing the papers read in these conferences and the papers specially written for these collections. The sum total of all this work has given rise to a much deeper awareness about the works and science-organizational activities of this great mathematician of the 18th century and of a number of his outstanding contemporaries, about the scientific contacts between the Academies of Sciences of Peterburg, Berlin and Paris, as well as those among the other scientific collectives and educational institutions. We should add here, that the publication of some of the volumes of series II of the Complete Collected Works of Euler, in particular of his works on mathematical physics (1960) and, on the theory of ship (1978), have also considerably enriched our knowledge of the mathematics of 18th century.

The number of new books on the history of mathematics of the 19th and 20th centuries considerably surpass all the above mentioned and related publications. In what follows, we shall be limiting ourselves to brief comments on some of the newer publications on the histories of mathematical analysis, theory of sets and, theory of functions of the real variable.

Thus, the manuscript entitled "Dissertations on the Theory of Mathematical Infinity", presented in 1785 by L. Carnot in the competition organised by the Academy of Sciences of Berlin, announced at the initiative of Lagrange, has been published (1970, 1979). At that time Lagrange was the Director of the Mathematics Class of the Academy and, he did not approve Carnot's manuscript, and that is why it remained unpublished. However, it is undoubtedly an interesting work. This was the first variant of Carnot's widely known "Meditations on the Metaphysics of Infinitesimal Calculus" (1797, 2nd ed. 1813). Carnot's basic idea—substantiation of analysis upon the principle of compensation of errors, dating back to G. Berkeley—is one and the same in the manuscript and in the published version, but the "Dissertation" contains some important moments anticipating Cauchy's reform: an understanding of the infinitesimal as a variable, of the connection between this concept and the concept of limit, and even an attempt at synthesizing the theory of limits together with its "strictness" and the infinitesimal calculus along with its algorithmic attainments. It is true that Carnot could not succeed here, and all this was successfully done by Cauchy. New light has also been thrown upon the so-called theory of limits of d'Alembert. This name does not quite exactly reflect the role of d'Alembert in the elaboration of the theory of limits: he was rather a successful propagandist of this theory, which was still in need of some important specifications and development; N. Bourbaki's evaluation of d'Alembert's definition of the limit as "very clear", is an overstatement. L'huilier's book—written under the influence of d'Alembert, but providing a broader treatment of the concept of limit, as has been noted by E.S. Shatunova (1966)—won the prize in the above mentioned competition.

Archival searches have also thrown a new light upon some aspects of the work of Cauchy. The beginning of the proof sheets of the second part of his famous course of lectures in the Polytechnical School has been found, and this has finally enabled us to date the proof of Cauchy's famous theorem about the existence of solutions for the system of first order differential equations, and at the same time to explain the reason impeding the publication of this second part, and namely the differences of Cauchy with the then more influential professors of the Polytechnic, on the question of level of teaching of analysis in the Polytechnic (C. Gilain, 1981). The materials available in the archives of Paris have also enabled us to specify the exact nature of M.V. Ostrogradsky's and V. Ya. Bunyakovsky's participation in the elaboration of Cauchy's theory of residues, at its early stage (1824-1826) and in its first applications in mechanics and in the theory of heat (investigations and publications of the present author and V.I. Antropova, 1965; V.S. Kirsanov, N.S. Ermolaeva, 1985). Here, one of the manuscripts of Ostrogradsky, presented at the Academy of Paris in the beginning of 1826, deserves special mention: herein Ostrogradsky's famous integral formula has been formulated and proved for the first time and thereby his priority has been finally established—it has at times been subjected to doubt; here we find the first generalisation of the method used by Fourier, during 1807-1822, for solving the problem of propagation of heat, wherein the ideas of D. Bernoulli and Euler were developed. Subsequently, the author of this work presented it in a somewhat revised form, to the Peterburg Academy of Sciences in 1828 and, it was published in the Transactions of the Academy in 1831. In 1827 Ostrogradsky submitted another memorandum to the Academy of Paris—on the propagation of heat in the right prism, having an isosceles right angled triangle as its base. He conveyed his solution to G. Lamé, who published his own enunciation of it in 1861, but the Russian translation of Ostrogradsky's memorandum was published only in 1965. Yet another valuable archival material, to some extent close to the one mentioned above, namely, the notes of Ostrogradsky's lectures on the theory of definite integrals, read in the years 1858-1859, in the hall of the Engineering Academy, has been published by V.I. Antropova (1961). Here many special integrals have been computed with the help of Cauchy's theory of residues and an originally enunciated theory of multiple integrals.

In this connection, here it must be mentioned, that in the 19th and early 20th centuries courses of differing volumes on the theory of definite integrals—computed this way or that, when the corresponding prototypes were not elementary functions or their superpositions—were read in many universities. P.L. Chebyshev read it for a number of years in the University of Peterburg, as an introductory course, together with the calculus of finite differences. Recently, N.S. Ermolaeva found the complete notes of the course on theory of probabilities, read by Chebyshev in the years 1876-1878, including both the introductory parts, and she read a paper on it in the International Congress of the Bernoulli Society. This paper is being published in the proceedings of the Congress, and the manuscripts of the said lectures are being readied for publication.

The place of honour in the elaboration of the foundations of the mathematical analysis of 19th century belongs to B. Bolzano, who largely anticipated Weierstrass, Dedekind and G. Cantor—both in his general conception and in a number of concrete results. His remarkable "Studies on the Functions" remained in manuscript form for nearly a hundred years, and was published by K. Rychlík only in 1930, and from among the works published in his lifetime, special mention must be made of the brochure, containing a "purely analytical" proof of the theorem about the intermediate values of a continuous function—which too failed to draw the attention of the leading mathematicians immediately. There was a gap in Bolzano's proof: the theory of real numbers was not enough for its completion. It has been found out comparatively recently, that evidently Bolzano himself noticed this gap, in any case the text of his elaboration of the theory of real numbers has been preserved; it predates the constructions proposed later on and independently of each other by

Weierstrass (1860), C. Méray (1869), G. Cantor and Dedekind (1872). K. Rychlík published this text in 1961; he is of the opinion, that Bolzano's theory, which is not quite clear and complete, may be brought up to the level of modern requirements of strictness, without substantial changes; on this all the specialists are not in agreement with him.

Bolzano was not only a predecessor of Weierstrass and Dedekind in the realm of ideas, but it appears—as has been shown by P. Dugac (1973)—that he influenced both of them. Weierstrass set forth his classical system of mathematical analysis, as well as the theory of analytical functions in his Berlin lectures. Weierstrass published some isolated new results contained in the lectures, but did not publish the course of lectures, as he was not satisfied with what he attained. Weierstrass' lectures were collected by a diverse international audience, but their contents were spread even wider, thanks to the printed versions prepared by some of his listeners. In 1973 P. Dugac published highly interesting notes of 4 courses of lectures read by Weierstrass during the years 1861-1887, and the corresponding correspondence of Weierstrass with a number of outstanding mathematicians. These materials substantially helped specifying the stages of development of the foundations of classical analysis of Weierstrass, as well as of his system of theory of analytical functions. Of no less importance is P. Dugac's book on R. Dedekind, published in 1974. Archival materials constitute half of this book. These materials are related to Dedekind, one way or the other; these are: his manuscripts, correspondence (in particular, with G. Cantor)—in part, not contained in the collection of their correspondence published by J. Cavaillès and E. Noether, letters of many other leading mathematicians etc. Dugac considerably specified the interrelationship of Cantor's and Dedekind's set theoretic ideas and the significance of their respective contributions to set theory, in the light of the subsequent logico-mathematical investigations right upto K. Gödel and P. Cohen. J. Dieudonné briefly formulated the general conclusion in his forward to this book by Dugac: Cantor's early work on countability, real numbers and topology remain his living and fundamental legacy, and in these fields Dedekind shares with him, in equal measure, the credit for laying the "set theoretic" foundations of the mathematics of our times. Unfortunately, P. Dugac failed to mention F.A. Medvedev's contributions in the elaboration of the history of set theory, who made a detailed and objective study of the connections between G. Cantor's and Dedekind's set theoretic investigations, in a book published in 1965. It is true, however, that he did not have at his disposal all the archival materials, brought into circulation by Dugac, who mentioned Medvedev's book only in passing. It is well known that set theory and theory of functions were developed vigorously in our country. The first shoots of the Moscow school of theory of functions, often called the school of D.F. Egorov and N.N. Luzin, date back to the beginning of this century. Hence, the interest in the life and work of these scholars and of their colleagues and followers, is quite understandable. F.A. Medvedev was the first (1959) to seriously study the given stage of the Moscow school. Subsequently, many archival materials, pertaining to the life and work of the pioneers of this school and of their followers, came to light: Egorov's letters to Klein, those of de la Vallée-

Poussin to Luzin, Egorov-Luzin and Luzin-Denjoy etc. correspondences, Luzin's preface to Euler-Goldbach correspondence, Luzin's opinions about the work of the famous geometer S.P. Finikov etc. Recently, new and important materials pertaining to the first formative years of the Moscow school of theory of functions have come to light; these are related to: the course of lectures on the theory of functions delivered by B.K. Mlodzievskii in the beginning of the 20th century, the papers read at the students' circle—where P.A. Florensky was one of the most active members and where Luzin look part, the discontinuous functions studied there etc. What is more, now new light has been thrown upon the role of N.V. Bugaev, whose philosophico-mathematical ideas clearly stimulated the interests of the younger scholars and students in the theory of functions. Herein also comes to light the connections with the Moscow school of philosophy. S.S. Demidov's, F.A. Medvedev's, A.N. Parshin's and other publications pertaining to these connections came to light during 1985-1986. In this connection one must mention the analysis made by the Colombian scholar L.C. Arboleda (1980) of the letters to M. Fréchet from P.S. Aleksandrov and P.S. Urysohn—who founded the Moscow school of topology in the beginning of the 20s of this century. Reminiscences of P.S. Alexandrov were published during the years 1971-1980. These are not archival materials, but we should not pass them over in silence, since here we find a brilliant description of mathematics in Moscow with the global developments of mathematics in the background, from the middle of the first decade of this century and for a stretch of more than sixty years. D.E. Meinshov's reminiscences of his student days and of the early years of his scientific career (1983), constitute a valuable supplement to the memoirs of Alexandrov.

Scientific correspondences were of first order significance during the 17th-18th centuries, when scientific periodicals were extremely weakly developed, scientists rarely met one another, lived in different towns and, scientific congresses and conferences did not take place at all. These correspondences did retain their place in scientific life even later on, right upto our times, though to a lesser extent and, to the examples just cited, here I shall mention only three: the correspondence of S.V. Kovalevskaya with G. Mittag-Leffler—edited by P. Ya. Kochina and E.P. Ozhigova—important, not only for the biography of this outstanding woman and mathematician, but also for investigations into the life of the international mathematical community during the 80s of the last century; the correspondence between V.A. Steklov and A. Kneser, in the beginning of the 20th century, on questions of mathematical physics and related themes—published by I.I. Markush and others (1980); and the correspondence of A.A. Markov with A.A. Chuprov, 1910-1917, on questions of probability theory and mathematical statistics, prepared for publication by Kh. O. Ondar (1977).

Conclusion. This survey of the historico-mathematical investigations of the last few decades, is far from complete; but even this survey shows, that these investigations have not only considerably supplemented our knowledge of the past of mathematics, but that they also entail considerable changes in our general notions about the characteristic traits of the developments of mathematics at different times and in different directions. This survey was divided into several points; in each of them a corresponding summing up has been provided with and, some of the open problems have been indicated. Now a few words remain to be said about some of the over-all changes in that retrospective, wherein the developments of the last four or four and a half thousand years of developments of mathematics had been presented until recently; it is about these years that we may speak sufficiently confidently. At issue here is the further specification of the periodization proposed by A.N. Kolmogorov [26]. It is true, that the periodization proposed here is global in character and, does not pretend to provide the universal characteristic traits of the objects and methods of mathematics at each of the indicated periods, thanks to the unevenness and inexact synchrony, and sometimes owing simply to the non-synchronous progress (at times regress) of mathematics in the different regions or sub-regions considered [26]; the same is true of the present survey. Having this stipulation in view, such global periodization of mathematics, understood as a single science, without its division into sub-disciplines, appeared to be fully satisfactory for a long time. Briefly speaking, A.N. Kolmogorov, made a distinction among four large periods:

1. Birth of mathematics, as it took place, for example, in Egypt, where, evidently, mathematical theory—in the sense of proofs of general theorems—did not exist at all, and everything was reduced to the collection of arithmetical and geometrical examples of practical importance, solvable at the level of simplest concepts, according to some prescriptions and rules for computation and measurement; however, perhaps, in Ancient Babylon the prescriptions were not confined to what was of direct practical necessity and, there arose more abstract scientific interests and they evolved some general algebraic methods for solving a number of problems (here, we have, almost word for word, reproduced A.N. Kolmogorov's expressions, from his article in the second edition of GSE, v. 26, 1954).

2. The elementary mathematics of the period 7th century B.C.-17th century A.D., when it became theoretic, having the constant magnitudes of arithmetical or geometrical nature as its main object; approaches to the ideas of an infinitesimal analysis are observed in Greece and then in medieval Europe, but they were not developed; this second period has been divided into two sub-periods in Kolmogorov's article entitled "Mathematics"—encompassing, correspondingly, the ancient (Greece, Hellenism, Rome) and the medieval periods (countries of the Orient and of Europe).

3. The period of emergence and development of the mathematics of variables, from the 17th century; it enters into [4.] the period of modern mathematics in the 19th century, in connection with extreme extension of the object of mathematics and the generalisation of its concepts and methods; A.N. Kolomogorov refrained from providing any direct global characterisation of this modem period. At the same time he has provided a fully intelligible account of the transition to modern mathematics, which began in the first decades of the 19th century, straight off in the two fields of geometry and algebra. His article began with Engels' definition of mathematics as the science of the quantitative relations and spatial forms of the real world; in the section devoted to modern mathematics Kolomogorov indicated

the range of "quantitative relations" and "spatial forms" studied, becomes extremely widened in this stage (this is explained with the help of some examples) and that, when these two expressions are so widely understood, even then, that is even at the present stage of development of mathematics, its initial definition holds good. Here Kolmogorov adds that, when the expression "quantitative relations" is interpreted sufficiently broadly, then "spatial forms" may be considered to be special kinds of "quantitative relations". Here it is not possible to enter into a discussion of the wide range of related methodological questions that arise, for example, of A.D. Alexandrov's treatment of geometry (1952) as the science of spatial relations and forms, as well as about the other relations and forms of reality, which are structurally similar to the spatial ones ("spatial-like"). B.A. Rozenfeld's proposal to generally call modern mathematics non-Euclidean, is hardly felicitous: this word is very closely connected with non-Euclidean geometry. This entire periodization is linked with the periodization of the prevailing social formations.

Many Soviet historians of mathematics accepted A.N. Kolomogorov's periodization with some modifications. Now, let me summarize the specific comments regarding the periodization already made; let me stipulate again, that the terminology used here is conventional and, the chronological and territorial boundaries are diffused.

1. While retaining the term "the period of birth of mathematics", we should stress the fact that not only in Ancient Babylon, but also in Egypt—the prescriptive form of enunciation was followed by works reflecting deductive arithmetical, geometrical and algebraic thought; true, not at the level of construction of systems like the Euclidean geometry. Perhaps, the expression "piecemeal deductive mathematics", more adequately describes the mathematical thought of the aforementioned and similar civilizations.

2. In Greece mathematics was transformed into a deductive science, in the Euclidean—if one may say so—sense, i.e., into a system of disciplines, axiomatically constructed at the level of Aristotelian logic; it was not accidental that the formation of this logic was almost simultaneous with the codification of Euclid's "Elements". However, this axiomatization did not spread to all the branches of mathematical knowledge and, in the 19th-20th centuries it has been substantially completed and developed. However, Greek mathematics is different from what preceded it, also in another, in principle more important, respect—important from the point of view of subsequent development of mathematics: in natural philosophy and in mathematics there arose the idea of infinity and it found application—it became the starting point of infinitesimal mathematics; it is in Greece that the fruitful interaction of philosophy and mathematics was established. Thus, the term "elementary mathematics" is hardly adequate for the content of Greek mathematics. Perhaps, here the more general, though less concrete, name of "the period of emergence of mathematics as a logico-deductive science", has some advantages.

3. Probably, the term "the period of elementary mathematics" is most appropriate, when one wants to provide a global characterization of the mathematics of the

middle ages. However, here it is important to have in view the essential traits of mathematics in Europe of the 14th-16th centuries and the progress of the non-elementary ideas and methods spoken above.

4. In reality, the 17th-18th centuries are characterised, first of all, by the primacy gained therein by the mathematics of the variable magnitudes. The problem of naming the period of modern mathematics in terms of its contents, happens to be more complex. Perhaps, it would have been proper to borrow a term from modern mathematics, and speak of this period as the period of "non-standard mathematics". However, the question arises: would it be correct to speak of the last two centuries, as one single period of development of mathematics? Does not the mathematics of the recent times of scientific and technological revolution, with its characteristically fast progress of informatics and of the other noticeable (discrete, finitary etc.) tendencies, constitute the first stage of a new period of development of this most ancient science? Personally for me, it is very difficult to provide an answer to-day.

Literature

1. *Montucla Jean-Étienne* Historie des mathématiques ... P., 1758; 2 éd. P., 1792-1802.
2. *Loria G.* Guida allo studio delta storia delle matematiche. See. ediz. Milano, 1946.
3. *May K.O.* Bibliography and Research Manual of the History of Mathematics. Toronto: Univ. Toronto Press. 1973.
4. *Cantor M.* Vorlesungen über Geschichte der Mathematik. Leipzig. 1880-1907. Bd. I-IV.
5. *Tseiten G.G.* Istoria matematiki v drevnosti i v sredine veka. 2-e izd. M.;L., 1938: *On zhe*. Istoria matematiki v XVI i XVII vekakh. 2-e izd M.;L., 1938,
6. *Klein F.* Elementary Mathematics from an Advanced Standpoint. Tr. E.R. Hedrick and C.A. Noble, 2 vols. New York: Macmillan, 1932 and 1939 [a Russian tr. of the same was published in 3 vols, in the years 1933-1935].
7. *Tropfke J.* Geschichte der Elementarmathematik. B,, 1921-1924. Bd. I-VII.
8. *Tropfke J.* Geschichte der Elementarmathematik. 4 Aufl. B.; N.Y. 1980. Bd. I.
9. *Cajorie F.* History of Elementary Mathematics with reference to the Methods of Teaching. 2nd Russ. ed. Odessa, 1917.
10. Enzyklopädie der mathematischen Wissenschaften, Leipzig, Bd. 1-6. 1898-1934; 2 Aufl. 1952-1968.
11. Enciclopedia delle matematiche elementari e complementi con estensione alle principali teorie analitiche e geometriche ... Milano, 1930-1950. vol. 1-3.
12. Problemy Gilberta (Hilbert's Problems). M. 1969.
13. *Klein F.* Lectures on the Development of Mathematics in the 19th Century. Russ. ed. M.; L., 1937. Part I.
14. *May K.O., Roebuck L.* World Directory of Historians of Mathematics. 2nd ed. Toronto. 1978.

15. Istoria matematiki s drevneischikh vremeon do nachala XIX stoletia. M., 1970-1972. T. 1-3; Matematika XIX veka: Mat. logika. Algebra. Teoria chisel. Teoria veroyatnostei. M., 1978; Matematika XIX veka: Geometria. Teoria analit. funktsii. M.. 1981.
16. Abrégé d'historic des mathématiques. 1700-1900/ Ed. J. Dieudonné. P., 1978. T 1-2.
17. Istoria otechestvennoi matematiki. Kiev, 1966-1970. T. 1-4.
18. Ocherki razvitia matematiki v SSSR. Kiev, 1983.
19. *Kline M.* Mathematical Thought from Ancient to Modern Times. N.Y., 1972.
20. *Bourbaki N.* Éléments d'historie des mathématiques . Russ. Ed. M., 1963.
21. *Struik D.J.* A Concise History of Mathematics. N.Y., 1967. [2nd Russ. ed. M., 1969.]
22. Dictionary of Scientific Biography. N.Y., 1970-1980. vol. I-XVI.
23. Scienziati e tecnologi dalle origini al 1875. Scienziati tecnologi contemporanei. Milano. 1975. vol. I-III.
24. *Waerden B.L van der.* Geometry and Algebra in Ancient Civilizations. B., 1983.
25. Scientific Change/ Ed. A.C. Crombie. L., 1963.
26. *Kolmogorov A.N.* Matematika // BSE. 2-e izd. 1954. T. 26.

Source: Zakonomermosti razvitia sovremennoi matematiki. "Nauka". M., 1987. s. 28-74.
About the Author: See, In memoriam: Adolf-Andrei Pavlovich Yushkevich (1906-1993), *Historia Mathematica*, Volume 22, Issue 2 (May 1995): 113-118.

Nonstandard analysis and the history of classical analysis

Fedor Andreevich Medvedev

The history of classical analysis is perhaps the most investigated part of the history of mathematics, and that is quite natural. Mathematical analysis has been viewed as the "simplest and most universal language ... most suitable for expressing the invariable relations of natural phenomena" [1, p. XXIII], That is why, the greatest representatives of mathematical thought, and the humblest workers in the field of mathematics, applied their strength to its construction as an important mathematical discipline, from the 17th to the 19th century, as well as, during the greater part of the present century. This new discipline, in the process of its construction, not only served as a felicitous means for describing the phenomena of the external world, but also provided a possibility for advancing profound philosophical reflections about the differential picture of the world, about the causal connections in it, about the laws of nature and thought. That is why, the historians of science have paid their greatest attention, to the history of analysis.

A major specificity of the approach of the historians to the study of the formation of this branch of mathematics has been, and to a considerable extent still is, to view it as a single integral theoretical discipline. To a certain extent this view reflects an aspect of a more general notion, according to which "mathematics grew as a single whole" [2, p. 13].

However, of late such a view about mathematics has been shaken or has at least been questioned. "The basic conclusion that may be drawn from the presence of several conflicting approaches to mathematics is as follows: there exists not one, but many mathematics" [3, p. 358]; not only some individual mathematicians and historians of mathematics, but even some philosophers share this conclusion [4, pp. 186-187]. An analogous hypothesis suggests itself also in respect of mathematical analysis: it began to take shape after the construction of the intuitionist and constructivist analyses, and became especially clear after the creation of nonstandard analysis, towards the middle of this century. Each of the systems mentioned, is sufficiently independent of, and definitely different from, any other of them—in terms of the composition of basic concepts, modes of advancing arguments about them and, computational procedures; and it is hardly probable, what is more, impossible (and even if possible, then not necessary) that they be united in a single theoretical construction. This especially comes to the fore, when we view analysis, not so much as a theoretical doctrine about its basic concepts, but rather as calculus, as formal system. Such an outlook on mathematical analysis has been considerably developed, at the level of historical studies, in a book [5] by C.H. Edwards. In so far as, "it is possible to propose a large number of formal systems, for describing one and the same fragment of reality" [6, p. 87], there is no ground for preferring one of them in advance: it would be expedient to use different calculi in different situations.

A distinctive and even strange specificity of theoretical constructions—and not merely of the mathematical ones—is this, that at a definite stage of their construction, and at times from almost the very beginning, the adherents of the corresponding theoretical schemas become tempted to view them as the universal and the only possible schema and to declare all the others as false. Such, in particular, was the situation in the history of mathematical analysis, in the second half of the 19th century, when this mathematical discipline attained the highest level of its development and when the gloss of Cauchy-Weierstrassian rigour was put on it. The analysis of Cauchy-Weierstrass was considered to be the only legitimate theoretico-analytical construction, and all that preceded it came to be regarded as mere approximations to it, and largely mistaken at that: "... in the textbooks of Cauchy at long last we are on firm ground" [7, p. 207].

The works on the history of analysis come out mainly as descriptions of the development of something integral, the one and only possible thing, which served as a kind of ideal for a single science, to which the constructions of the previous epochs approximated. From the point of view of this ideal science (which is in practice reducible to a construction, put forward in some standard textbook, followed by the author of the corresponding "history of analysis" or part thereof), the problems that emerged in the course of history had singular correct solutions; and it is these that were mainly of interest to the historians. Indeed the approaches to them could be different: to a large extent these are unsatisfactory and even patently wrong, and then, in the course of further searches they are usually rejected, or at best they are used as auxiliary devices, serving as the raw material, from which the ideal science was built, or they provide heuristic indications, which lead to the discovery of the absolute truths of the ideal science.

In particular, the "evolution of "rigour" in analysis can be summed up as a continuous ascent from unclear and vague concepts, their gradual elucidation, and then the arrival of a stage of stability—after which it was impossible to have *any* dispute regarding what constitutes a correct proof in analysis" [8, p. 50]. In the opinion of J. Dieudonné, the analysis of Cauchy-Weierstrass happened to be this stage of stability.

The analysis of Cauchy-Weierstrass took shape in the 19th century. However, before that mathematical analysis went through more than two thousand years of development—from the first quadratures and cubatures of the ancient Greeks to the analysis of Newton-Leibniz, crowned with the works of J.L. Lagrange and L. Euler. Here we shall mention only some of the landmarks along this road.

It is well known, that in ancient Greece the first quadratures and cubatures were carried out with the help of some infinitesimal procedures [9]. These procedures were so fruitful, that even after the elaboration of the method of exhaustion by Eudoxus, which tended to exclude the infinitesimals from mathematical arguments and soon became the official doctrine on the corresponding questions, infinitesimal considerations continued to be in wide use, and among the users there were adherents of the method of exhaustion; the assertion of S. Ya. Lurie, that if the followers of the method of exhaustion "did not get hold of ready-made solutions, already discovered by the atomists, then they themselves preliminarily found them out—by stealthily applying [the method of] atomistic decomposition" [9, p. 159], is best illustrated by the example of Archimedes.

With the renewal of interest in the problems of analysis in the 17th century, infinitesimal considerations again came to the fore (in the works of J. Kepler, B. Cavalieri, G.P. Roberval and of many others), though the spell of the method of exhaustion also continued. It was considered to be an irreproachable, totally rigorous mode of mathematical reasoning in respect of the problems of quadrature, cubature, centres of gravity, maximums and minimums, and tangents; and mathematicians did not see in it those serious logical flaws, which may be observed today; for example, this, that therein, neither were the magnitudes under consideration defined, nor were their existence proved. Though this spell was retained for a long time, right upto A. Cauchy and even later, nevertheless the method of exhaustion more and more retreated to a secondary position. It was substituted by the *infinitesimal calculus*.

Namely, thus was created the grandiose house of mathematical analysis—the most extensive and the most fruitful mathematical discipline.

Almost from the very beginning of its construction, strange reproaches began to be sounded regarding the illogicality of the arguments, involving the infinitesimals, provided in this calculus—these reproaches continue even to-day [3, pp. 151-177]. However, much more illogical was the very demand, that the modes of reasoning used in analysis are to be subordinated to the norms of existing logic. Mathematical analysis is based on the concept of function—which is a special instance of relation (and for an extended understanding of function, they are one and the same). Logic of relations did not exist earlier and it was constructed only in the second half of the 19th century. [It is true that individual advances were made in that direction. Such attempts date back to Aristotle,* G.W. Leibniz, J.H. Lambert and to some others, but the construction of the theory of relations as a section of logic began only with A. De Morgan, or even later—with C.S. Peirce.] That is why, the arguments employed by the mathematicians in analysis, naturally, did not fit into the framework of the arguments conducted in accordance with the canons of older thought. Mathematicians did not like to wait till the emergence of the corresponding logic, in fact they paved the path for it with their new calculus, having constructed a formal system, which is a special instance of a large fragment of the logic of relations (this was done by Euler, Lagrange and their followers).

In the final analysis this formal system was infinitesimal, and Lagrange's heroic attempt at liberating mathematical analysis from the infinitesimals turned out to be totally unsuccessful, in spite of his indisputable achievements. [In J. Grabiner [10] one finds a good description of many achievements of Lagrange.] What is more, even Cauchy's reconstruction of analysis carried the mark of that very infinitesimal. On this, one must dwell in greater detail.

In correspondence with the established tradition, earlier we have united the constructions of Cauchy and Weierstrass into an aggregate. Indeed, there is much in common in their approaches, which provides a definite basis for such unification. However, in essence the constructions of Cauchy and Weierstrass, differ by so many fundamental parameters, that it would have been more correct to speak of two different systems of analysis, created by them. This difference has been quite clearly outlined by N.N. Luzin [11, pp. 305-312], and in his words, generally speaking, this difference lies in the fact that, Cauchy "in principle, introduces variables, and, thus, since then analysis stands enriched by these new magnitudes, used equally rightfully with the constants, just as the imaginary numbers are used on a par with the real ones" [11, p. 305]; Weierstrass, however, "first of all removes all the variables from analysis, any change, motion and everything is reduced to the stationary conditions and to that alone, i.e., to the constant magnitudes" [11, p. 307]. These systems are also different in terms of the explications of the concepts of infinity, which are foundational to them. Cauchy based himself upon the concept of potential infinity and, the concept of potentially infinitely small magnitudes are central to his system

* To: Gaṅgeśopādhyāya, c. late 12th century, India – Ed.

of analysis, whereas Weierstrass leaned on the concept of actual infinity, and he in fact drove away the infinitesimals from analysis. This is not the place for citing the other differences, many of them are yet to be brought to light; here we shall only mention the fact that Cauchy's conception completed the stage of infinitesimal analysis, whereas in Weierstrass' system, analysis was restructured as a set theoretic discipline; its contours have been quite clearly outlined by Luzin [11, pp. 307-312].

Infinitesimal considerations were put aside in the classical set theory, and the actual infinitesimals were denied every right to exist; G. Cantor fought against them energetically—on this, see, for example, [12, pp. 294-296]. At the end of the 19th and during the first half of the 20th century, the Weierstrassian system of analysis, based on the theory of sets, became the ideal of analytical construction—it was considered to be singularly legitimate and absolutely true. Armed with it, even the historians of analysis started viewing the previous developments through the prism of such construction. Naturally, there arose all possible distortions. These are quite numerous, and there is no scope here, even for providing a simple list of them. We shall cite only a few examples.

After a very brief and not entirely objective description of B. Cavalieri's methods of quadrature and cubature, O. Toeplitz termed his conclusions—"clever" [13, pp. 55-58], but there itself he characterized them as "untrustworthy to the highest degree". After that he admitted, that in the 17th century the doctrine of indivisibles was "lauded to the skies" and was employed in various forms by P. Guldin, B. Pascal and J. Kepler and, that in the works of G.W. Leibniz the various trends in the use of the infinitesimal methods came together, "as rays converge in the focus of a lens" and illumined the entire 18th century. Toeplitz concludes that L. Euler, D. Bernoulli, B. Taylor and others "constructed the new edifice of mathematics basing themselves upon such non-strict and heuristic methods" [13, p. 59]. Thus, it appears that a large mathematical theory was built upon the precarious foundations of doubtful heuristic methods, and only a "clever instinct" saved Cavalieri and, of course, the other mathematicians from making false moves.

While describing the mathematical analysis of the 18th century, J. Dieudonné, on the one hand, calls Euler one of the two "giants of that century" [14, p. 20] (the other one being Lagrange), and on the other—presents him as nearly the greatest fumbler: he did not define the concept of a "continuous" or "regular" function "more precisely"; did not give an "exact definition" of his "mechanical functions"; his conception of numerical serieses was "shaky and imprecise"; he "could not formulate" the definition of the concept of the sum of a series, basing it on the concept of limit, and though "he knew well, that when the general term of the series

$$a_0 + a_1 + \dots + a_n + \dots,$$

does not tend to zero, then one should not speak of the "convergence" of this series in the usual sense, nevertheless he thought that it was often possible to calculate the "sum" of this series" [14, pp. 21-22]; Euler "did not make a clear statement" about one of the divergent serieses; "Euler's interpretations of the different meanings of the word "sum" of a series appear to be very confused, leading to contradictions",

"Euler did not at all understand the difficulty inherent in his definition of the "sum" of a series" [14, p. 23]; Euler "did not always remember what he wrote a few years back", "he did not elucidate the phenomenon of non-uniform convergence, when he ran into it" [14, p. 30], etc.

In short, it appears, that only with Cauchy did the mathematicians—to use an expression of Bourbaki, indicated above—find themselves "on a firm ground", and till then they wallowed "in the quagmire of mathematical analysis" of the Newtonian-Leibnizian-Eulerian kind [3, p. 151], and Cauchy too, quite often fell into it.

Demolition of such mistaken ideas about the development of mathematical analysis began after the construction of the so-called non-standard analysis, in the middle of the present century. [The construction of this analysis is to a great extent connected with the name of the American mathematician A. Robinson [1918-1975], and his work published in 1961 [15] is considered to be the first work in the field. This is not entirely correct. Even if we refrain from mentioning its pre-history, which dates back to the 19th and the first half of the 20th century, we may mention, for instance, the work of C. Schmieden and D. Laugwitz. published in 1958 [16] and the papers of the latter: [17], [18]. A.F. Monna began working in this direction [19, p. 4], since 1950.] The sole fact, that the infinitesimally small magnitudes and numbers, that were assiduously driven away to build the Weierstrassian analysis, again got back their right to citizenship in mathematics, itself compels us to re-examine the ideas so far formed, about the course of development of analysis upto Cauchy and Weierstrass, about the character of its infinitesimal apparatus and, about the modes of reasoning employed therein. Now the infinitesimal procedures are no more considered to be "non-strict, heuristic methods", but are rather viewed as quite honest means of advancing mathematical reasoning, which not only enable us to re-state the well-known results with great simplicity and elegance, but also help us discover new results; as an example of the latter, one may mention the so-called solution-configurations for a determinate class of ordinary differential equations [20].

This re-examination still continues. It began with the efforts of the founders of non-standard analysis—Laugwitz (1965) and Robinson (1966) [21 and 22], Now the front in favour of such re-examination is quite broad, and the mere listing of the works in the already indicated direction, would take a lot of space. In particular, many results of Euler, so far considered to be "non-strict" and to have been obtained by doubtful means, turned out to be rationally interpretable within the frame-work of these new ideas [23 and 24]. Now they are no more the products of some special and secret intuition, but are provided with a rational reconstruction in a different system of analytical thought. Now there has even arisen a danger, of too direct an interpretation of some isolated notions of Leibniz, Euler and of some other mathematicians, of their unnecessary modernization in the spirit of non-standard analysis. W. Felscher has characterized this danger quite clearly [25, pp. 179-180]. However, one may state—by generalizing, to some extent, Laugwitz's understanding of some of the results of Euler [23, p. 4]—that, we must examine the attainments of our predecessors from the modern points of view, while keeping in mind the fact that they could not have thought, namely, in the way we do; all the same, we can rationally interpret what they did, namely, with the help of modern ideas.

Further, one of the basic principles of non-standard analysis is the principle of transfer/translation, according to which every true statement of classical analysts (the classical theory of sets) is true in the non-standard analysis (non-standard theory of sets) too, and conversely [see: for example, [20, p. 81]; and see: [26, pp. 202-203] for a more detailed formulation]. Thus, non-standard analysis appears to be yet another model for that very class of analytical phenomena and relations among them, for which the model of classical analysis was built. It is understandable, that in an atmosphere of supremacy of the classical model, the latter played the role of a paradigm in the sense of T. Kuhn or, that of a research programme in the sense of I. Lakatos, and the historians of analysis examined the results obtained by their predecessors through the prism of the basic propositions of this paradigm or programme. In particular, since in it [the classical model] there was no room for the actually infinitesimally small, so all manipulations with them, met with by the historians, were viewed with suspicion, and the data obtained with their help, were interpreted as to have been found non-rationally—intuitively or through the mediation of unreliable heuristic methods. This led to the, not quite correct, prevalent interpretation of historico-scientific facts, herein indicated (and largely, not even indicated).

After the construction of non-standard analysis, classical analysis of the 19th century ceased to be the only correct analytical construction, its universalization and absolutization appeared to be illegitimate forms of activities. At least two formal systems possessing equal rights turned out to be equally suitable "for describing one and the same fragment of reality". In its turn, the notions about some mythical "absolute strictness"—turned out to be a myth. That is why there arose the necessity of substantial corrections in the historiography of mathematical analysis—corrections, that have been made complicated owing to the presence of intuitionist and constructivist analyses.

Everything said and done, non-standard analysis permits us to give an answer to a ticklish question, which arose in connection with the earlier approaches to the history of classical analysis: if we admit that the infinitely small and infinitely large magnitudes are contradictory concepts, then how could the grandiose edifice of one of the most important mathematical disciplines be built upon them? H.J.M. Bos—one of the keenest of modern historians of analysis—has been compelled, inspite of certain prejudices against the non-standard analysis, to declare that "the recently constructed non-standard analysis provides an explanation, as to why analysis could develop upon the precarious foundations of the infinitely small and infinitely large magnitudes" [27, p. 13].

Literature

1. *Fourier J-B.J.*, Théoric analytique de la chaleur (1822) // Oeuvres. T.1. P., 1888.

2. Istoriya matematiki s drevneischikh vremeon do nachala XIX stoletiya. M., 1972. T.3.
3. *Kline M.*, Mathematics: The Loss of Certainty. New York. 1980. [Russ. Tr.: Moscow. 1984].
4. *Panov M.I.*, Metodologicheskie problemy intuitsionistskoi matematiki. M., 1984.
5. *Edwards C.H.*, The Historical Development of the Calculus. N.Y., 1979.
6. *Ershov Yu.L.*, Nekotorye voprosy primeneniya formalizovannykh yazykov dlya issledovoniya filosofskikh problem // Metodol. problemy matematiki. Novosibirsk, 1979.
7. *Bourbaki N.*, Elements d'historie des mathématiques. P., (Russ. Tr.: M., 1963].
8. *Dieudonné J.*, La notion de rigueur en mathématiques // Conference prononcé le 8 décembre 1982 au Séminar loi dé philosophic et mathématiques. P., 1983. Reprint.
9. *Lurie S. Ya.*, Teoriya beskonechno malykh u drevnykh atomistov, M.; L., 1935.
10. *Grabiner J.V.*, The Origins of Cauchy's Rigorous Calculus. Cambridge; L., 1981.
11. *Luzin N.N.*, Differentsialnoe ischislenie // BSE 1-e izd. 1934. T.22; *On zhe.* Sobr. soch. M., 1959. T.3.
12. *Kantor G.* (*Cantor G.*). Trudy po teorii mnozhestov. M., 1985.
13. *Toeplitz O.*, Entwicklung der Infinitesimalrechnung. B., 1949.
14. *Dieudonné J.*, L'analyse mathématique au dix-huitième siècle // Abérgé d'historie des mathématiques. 1700-1900. Algebra, analyse classique, théoric des nombres. P., 1978.
15. *Robinson A.*, Non-standard Analysis // *Konikl. Akad. Wetensch. Amsterdam. Ser. A.* 1961, Bd. 64, N 4.
16. *Schmieden C. Laugwitz D.*, Eine Erweiterung der Infinitesimalrechnung // *Math. Ztschr.* 1958. Bd. 69.
17. *Laugwitz D.*, Anwendungen unendlich kleiner Zahlen, I: Zur Theorie der Distributionen // *J. reine und angew. Math.* 1961. Bd. 207.
18. *Laugwitz D.*, Anwendungen unendlich kleiner Zahlen, II: Ein Zugang zur Operationsrechnung von Mikusinski // Ibid. Bd. 208.
19. *Bertin E.*, *Blij F.*, *van der*, *Hornix E.*, Curriculum vitae de A.F. Monna // Symposium dédié à A.F. Monna. Utrecht, 18 déc. 1979 // *Communications of the Math. Inst.* Utrecht, 1980. Vol. 12.
20. *Zvonkin A.K.*, *Shubin M.A.*, Nestandartnyi analiz i singulyarnye vozmuscheniya obyknovennykh differentsialnykh uravnenii // *Uspekhi mat. nauk.* 1984. T.39. vyp. 2.
21. *Laugwitz D.*, Bemerkungen zu Bolzano Grössenlehre // *Arch. Hist. Exact. Sci.* 1965. Bd. 2, N 5.

22. *Robinson A.*, Non-standard Analysis. Amsterdam. 1966.
23. *Laugwitz D.*, Eulers Begründung der Analysis aus der Algebra. 1983. Prepar. N 728.
24. *Laugwitz D.*, Fall and Resurrection of Infinitesimals // Textes UMFUFA, 1984. N 88.
25. *Felscher W.*, Naive Mengen und abstrakte Zahlen, II: Algebraische und relle Zahlen. Mannheim etc. 1978.
26. *Stroyan K.D.*, Infinitezimalnyi analiz krivykh i poverkhnostei // Spravochnaya kniga po matematicheskoi logike: Teoriya modelei. M., 1982. Ch. 1.
27. *Bos H.J.M.*, Differentials, Higher Order Differentials and the Derivative in the Leibnizian Calculus // *Arch. Hist. Exact. Sci.* 1974. Bd. 14. N 1.

Source: Zakonomernosti razvitiya sovremennoi matematiki. "Nauka". M., 1987; s. 75-84.
About the author: See, In memoriam: Fedor Andreevich Medvedev (1923-1993), *Historia Mathematica*, Volume 22, Issue 1 (February 1995): 88-92.

The new structural approach in mathematics and some of its methodological problems

Georgi Ivanovich Ruzavin

The structural approach in mathematics has found its most complete development in the works of a group of French mathematicians writing under the pseudonym of N. Bourbaki, and of late in the works of Mac Lane and, of the other scholars engaged in the elaboration of the algebraic theory of categories. This approach offers an opportunity to take a new look at many major problems of philosophy and methodology of mathematics. The most important among these problems are: the specificity of the object and method of mathematics, the place of mathematics in the system of scientific knowledge and, the relation of mathematical structures with objective reality.

The general problems of the application of mathematics in the other sciences and in practical activities become more clear, when they are viewed from the point of view of abstract structures. The question of the interrelationship of "pure" (theoretical) mathematics with applied mathematics simultaneously finds a more thoroughgoing solution.

1. Formation of the concept of abstract structure and the emergence of a new approach to mathematics

Progress in mathematics has always been connected with the growth in the abstractness of its concepts and theories. Modern mathematics uses ever deeper abstractions to study, not only the quantitative, but also the more complex structural relations, the traditional quantitative relations among magnitudes happen to be a constituent part of these more complex relations.

The beginning of this new approach to the subject-matter of mathematical investigations is, to a considerable extent, connected with the discovery of non-Euclidean geometry by N.I. Lobachevsky and J. Bolyai. It is difficult to overestimate the general scientific and theoretico-cognitive significance of this discovery. It not only undermined the centuries old belief in the possibility of one and only, one geometry of Euclid, but also fundamentally changed our earlier notions about geometry, about mathematics as a whole. Above all, the theoretico-cognitive lesson offered by the discovery of the non-Euclidean geometries consisted in the following: it convincingly demonstrated that the axioms of geometry are neither empirical descriptions and inductive generalizations of the properties and relations of the real physical space, nor are they *a priori* synthetic judgements—as I. Kant thought them to be.

Mathematicians, following B. Riemann, began to view these axioms as hypotheses of a kind, whose applicability to the study of the surrounding space must be established after providing suitable interpretations for the basic concepts of a geometry. Within the framework of mathematical investigations these concepts (of a point, straight line and plane) themselves remain as abstract as, say, the algebraic formulae. No one doubted the fact that the symbols of these formulae may indicate any number, and subsequently any vector, matrix, function or other object. However, for a long time the statements of geometry remained associated only with the properties and relations of physical space. It is that is why, namely, that the discovery of the non-Euclidean geometries had such a decisive significance for the emergence of a new approach to the subject-matter of geometrical investigations.

In so far as axioms can describe the properties and relations of objects having the most diverse concrete contents, we cannot pass judgements about their truth or falsity, basing ourselves upon any one system of objects, serving as their interpretation. Besides, intuitive self-evidence also cannot serve as a criterion for truth, since that which is self-evident to one, may not appear to be self-evident to another. That is why, the demand that the axioms should be intuitively self-evident is not a mathematical, but rather a psychological demand. From the logico-mathematical point of view, the most important criterion, which must be satisfied by any system of axioms, is the simultaneity or formal non-contradictority of the system of axioms.

If a system of axioms is contradictory, it will not have any interpretation and, consequently, will have no scientific worth whatsoever. The demands for completeness and independence of a system of axioms are not that obligatory, if only owing to the fact that a dependent axiom can always be translated into a class of theorems, and the criterion of completeness is applicable only to the comparatively simple axiomatic systems.

The transition from the concrete, contentful axiomatics like, for example, the axiomatics of the Euclidean elementary geometry, to the abstract axiomatics like the axiomatics of Hilbert, and then on to those fully formalized axiomatics, wherein symbols substitute terms, and propositions are transformed into formulae—is quite a new step in the development of the axiomatic method; it is sometimes called a revolution in this method. As we have noted above, it is, namely, this approach which provides an opportunity for viewing axioms as abstract forms; these forms may be used for investigating the properties and relations of various things that differ in their concrete contents.

In the formation of the ideas about abstract structures, a significant role has been played by set theory, which emerged towards the end of the 70s of the last century. This theory was crafted, in the main, by the works of the great German mathematician G. Cantor, directed at providing a satisfactory foundation to classical mathematics. This theory views the objects of all mathematical theories like the number, function, vector, matrix etc., in isolation from their mathematical contents. For the set theory these are but elements of such infinite sets, which may be handled with definite rules. Such an extremely general and abstract approach provided an opportunity for viewing the subject-matter of the most diverse mathematical disciplines from a single point of view. Namely, that is why, with the passage of time the set theory came to be looked upon as the foundation of the entire house of classical mathematics.

In the end, a synthesis of the set-theoretic ideas and the axiomatic method led to a new conception of the abstract mathematical structure. This conception was floated towards the beginning of this century. "One is tempted to acknowledge that the modern concept of "structure" was in the main formed around the year 1900; actually, it took another thirty years of study to make it fully clear"—wrote N. Bourbaki [5, s. 33]. This work was done by that talented team of French mathematicians, which, uses the collective pseudonym of N. Bourbaki. They characterize a structure as follows: "In order to define a structure, one or more relations, containing their elements are at first specified ... then it is postulated that the given relation or relations satisfy certain conditions (these are listed and they constitute the axioms about the structure under consideration). To construct an axiomatic theory of a given structure, is to deduce logical conclusions from the axioms about that structure, without admitting any other presupposition in respect of the elements under consideration (in particular, staying clear of any kind of hypothesis regarding their "nature" [5, s. 251.]

From this, it is evident that concrete interpretations of the objects of mathematical investigations are quite unimportant for the said investigations. One can view these objects as elements of abstract mathematical structures, all the essential properties of which are specified with the help of axioms. N. Bourbaki have classified the mathematical structures into three basic types, in accordance with the character of these properties; these are: the algebraic, order, and topological structures, More complex or multiple structures are formed by combining these initial structures. The structure of various mathematical theories may be studied and thereby they may be classified, with the help of these more complex structures. If earlier mathematical theories were simply elaborated one beside the other, in the course of their historical emergence, then now it is possible to reveal their deeper resemblance. Thus, for example, the set of natural numbers, which serves as the basis for the construction of mathematical analysis, contains all the three generative structures and hence, the earlier isolation of algebra, geometry and analysis turns out to be groundless.

Of late the algebraic theory of categories is drawing ever greater attention of the specialists. Before the emergence of this theory, the set was considered to be the most general concept of mathematics. That is why, while substantiating mathematics, all other mathematical concepts were sought to be defined by using the terminology of set theory. Even the conception of N. Bourbaki was no exception to this, based as it is, in the final count, on an axiomatic theory of sets. The concept of category is not only an alternative to the concept of set, it is also the further concretization and development of the idea of mathematical structure.

A category is made up of a certain class of objects and a definite class of morphisms, wherein each ordered pair of objects is contrasted with a corresponding set of morphisms. An operation with the morphisms, with the help of which from two given morphisms of a category, a third unique element is found out from among the set of morphisms, is called the composition or production of morphisms, It must satisfy the conditions of associativity and identity [4, s, 9-11]. Simply speaking, in the theory of categories the sum total of objects are considered together with their structure and with some representations among them, retaining the given Structure. If sets are viewed as the objects of a category, and representations among them serve as morphisms, then we get the category of sets. Thus, the concept of set turns out to be a special case of the more general concept of category. In contrast to the static concept of a set, here the principal attention is turned towards the character of representations, which retain the definite structural specificity of the objects, and thereby the active, constructive aspect of mathematical knowledge is underlined.

2. The role of abstract structures and categories in modern understanding of the subject matter and method of mathematics

The ideas and methods of the theories of structure and category provide a better opportunity to understand the qualitative change, which is taking place in the very subject-matter of mathematics, as well as in the application of its methods in the other sciences.

Upto almost the middle of the last century mathematics was viewed as a science of magnitudes and spatial figures. Of course, therein not the concrete physical, chemical etc. properties of the magnitudes, but the properties and relations common to all of them were taken into consideration. Since every magnitude can be expressed numerically with the help of a suitably chosen unit of measurement, in the past, often the essence of mathematics was considered to be located in the investigations regarding the properties of and dependencies among numbers [7, s. 15]. The study of spatial figures in geometry was also, in the main, limited to their metric properties. Though by the middle of the last century, there did exist in mathematics such theories and separate disciplines, wherein the questions of measurement did not play any important role (for example, in projective geometry, group theory etc.), nevertheless, the view that mathematics is a science about the metric properties of and relations among magnitudes, was dominant among mathematicians.

With the emergence of the new abstract divisions of mathematics, a structural approach towards the objects of mathematical investigations took shape; it became ever more clear that the subject-matter of mathematics is not limited to the study of the properties and relations obtainable among magnitudes and spatial figures. One of the important methodological conclusions, emanating from these latest results of mathematics, is this that notwithstanding the practical significance of the metric relations among magnitudes and their representations in numbers and functions, in the theoretical sense, they constitute only a part of the more extensive and deep-going teachings about mathematical structures and categories.

In their attempts at underlining the difference between the modern and the classical mathematics, some scholars often view the modern as the "qualitative" and the classical as the "quantitative" mathematics. However, such a contraposition is, in essence, based on an identification of the concepts of magnitude and number with quantity, and of the abstract structures and categories—with quality. One cannot agree with this position. It is understandable that nobody will object to the position that the concepts of structure and category are qualitatively different from the magnitudes or figures of the three dimensional space. There is, also, no denying the fact that modern mathematics has raised the investigations about the real world to a qualitatively new height. However, this does not mean that now mathematics has gone over to the study of the qualitative specificities of objects and processes. It is evident that by contrapositing modern mathematics to the classical, the deeper and more abstract character of the concepts and theories of modern mathematics are sought to be highlighted, the broadening of its scope and sphere of application is underlined—but by no means a transition to the qualitative methods of investigation is indicated. In contrast to the methods of the concrete natural and social sciences, the methods of the theories of structures and categories are mathematical methods, and not the methods of special sciences, which include observation and experiment. Had the converse been true, mathematics would have been then turned into a branch of the natural sciences and thereby it would have lost its character as a "generally significant and abstract science", to which F. Engels drew our attention.

The concepts and methods of the theories of structures and categories have made the process of application of mathematics in the other sciences, technology and practical activities considerably easy. Actually, having proposed the suitable abstract structures, the scholar or the practical worker can limit oneself only to verifying whether the objects under investigation satisfy the axioms of the structures under consideration or not. The entire further tedious and difficult work of deducing the conclusions from them becomes unnecessary, since one can straight off use all those theorems which were obtained while studying the mathematical structure considered. Thus, the abstract structures and categories of mathematics may be compared with ready-made forms, that may be used while investigating the phenomena and processes having the most diverse contents.

Abstract structures can be successfully used for constructing mathematical models; we may especially use those among them, which aim at revealing not only the numerical (metric) dependencies among magnitudes, but also the relations of a non-metric character. The study of such non-metric relations is of considerable significance for those sciences, where owing to the complexity of the object under investigation, and sometimes also owing to the unelaborated stage of a theory, it is impossible to present the results numerically. That is why there one is often required to turn to the abstract structures of order. In their investigations about the different types of relations obtainable among individuals and groups in social collectives, psychologists and sociologists have begun to apply the oriented theory of graphs, which constitutes the simplest form of algebraic category.

The experience of applying the latest structural methods in the exact natural sciences convincingly shows the future possibilities open for this line of mathematisation of the sciences, in fact, the use of the abstract structures of mathematics in such branches of the exact natural sciences as the theory of relativity and the quantum mechanics, theory of elementary particles and cosmology, quantum chemistry and molecular biology etc., is dictated by the very level of development of these disciplines. The concepts and theories of these disciplines, not only very often do not permit visual representations, but also do not admit of their description in the language of classical mathematics. That is why, there one is required to turn to the ideas and methods of the abstract structures and categories of modern mathematics. Thus, these abstract structures go to highlight the remarkable idea of V.I. Lenin regarding the fact that scientific abstractions, laws and theories do not push us away from objective truth, but rather take us closer to it. "Thought proceeding from the concrete to the abstract—provided it is *correct*... does not gel away *from* the truth but comes closer to it. The abstraction of *matter*, of a *law* of nature, the abstraction of *value* etc., in short *all* scientific (correct, serious, not absurd) abstractions reflect nature more deeply, truly and *completely*" [3, s. 152; Eng. ed., v. 38, P. 171].

Yet another important role of the mathematical structures lies in the fact that they serve as an exact language for the abstract description of the most diverse phenomena and processes. It is enough to note the fruitfulness of the method of mathematical hypothesis in the process of formation of the quantum theory in physics. In this connection Dyson wrote: "For physics, mathematics is not only an instrument, with the help of which it can quantitatively describe any phenomenon, but is also an important source of such ideas and principles, on the basis of which new theories are generated" [6, s. 112]. This specificity of modern mathematics, as an exact language, for abstractly describing the interconnections among phenomena, characterizes its role as a synthesizer in the general process of development of scientific knowledge.

While discussing the question of inter-object interconnections, it is necessary, in the first place, to turn our attention, to the role of mathematics, namely, as an exact language for expressing the dependencies, that we come across in physics, astronomy, chemistry and in the other branches of natural science. The mathematical language of formulae, equations and functions allows us to express the interrelations and laws of the phenomena investigated in every special science, in the most exact and general form. However, for finding out the adequate mathematical language one must take the specific, qualitative character of these phenomena into consideration. All these go to show, that in the real practice of scientific cognition, there exists a dialectical interconnection and reciprocity between the quantitative mathematical methods and the qualitative methods characteristic of every special science. The better we know of the qualitative specificities of the phenomena, the more successful we become in using the quantitative and mathematical methods, for analysing them. The establishment of the quantitative regularity of phenomena, is always based upon the ability to reveal that which is similar and common in what is inherent to the qualitatively different phenomena. This is possible only by studying the phenomena within the frame-work of the special disciplines. The entire powerful apparatus of mathematics turns out to be effective only in that case, when that which is similar and common in the phenomena under investigation, is preliminarily discovered and formulated in the form of sufficiently deep going general concepts and qualitative dependencies. In fact, if this or that science proposes only the simplest of inductive generalizations of facts and empirical laws, wherein the connections among those magnitudes are established, that are immediately observable in the experiments, then it is impossible to count on the application of the latest methods of mathematics, for their quantitative analysis. The history of physics, chemistry, astronomy and of the other sciences clearly testify to the fact that the progress of theoretical investigations in them was accompanied by an extensive application of the mathematical methods. Often the demand for elaboration of these theories promoted the emergence of new mathematical methods, a clear example of this is the emergence of infinitesimal analysis.

3. The mathematical structures and the real world

Just as the question of the relation of consciousness and being, is basic for philosophy as a whole, likewise the question of the relation of the mathematical structures and the real world happens to be central for philosophy of mathematics. In contrast to the positivist approach, the school of N. Bourbaki, not only does not ignore this problem, but, on the contrary underlines the fact that, "the interrelationship of the universe of experiment with that of mathematics" is the basic philosophical problem [5, s. 258]. This school also does not deny the existence of close ties among the structures of mathematics and empirical reality, though it considers the reasons of their existence to be entirely unexplainable.

The difficulties connected with the understanding of the objective nature of the abstract structures of mathematics, are rooted in the very specificities of mathematical knowledge, which bases all its propositions upon the laws and principles of logic and not on experiment. It was not accidental, that is why, for G. Frege, B. Russell and their followers to have attempted to seek the foundations of the entire pure mathematics in logic. However, therein they ignored the doubtless fact that mathematics, as an independent science, needs its own initial concepts and postulates. Otherwise, as has been correctly pointed out by H. Poincaré, it would turn into a grandiose tautology. It is also important to pay attention to the fact that many western scholars consider logic itself to be an a-priori-science about the forms of thought; and in so far as logic plays the most important role in the formation of abstract structures, often these structures are themselves also viewed as a-priori-forms.

The strictly logical and deductive character of the constructions and substantiations of mathematics are, ultimately, determined by the specificities inherent to the processes of abstraction and idealization in mathematics. Firstly, in mathematics abstraction proceeds significantly further than, say, in the natural sciences. In the concepts of the geometrical point, line, the variable, the function and in the case of all the other mathematical concepts in general, we abstract from the concrete contents and qualitative specificities of objects and processes. Secondly, many abstractions of modern mathematics emerge through a series of successive stages of abstraction and subsequent generalization. It is, namely, thus that all the mathematical structures have been formed. Thirdly, the relative independence of the purely theoretical development is, perhaps, more characteristic of mathematics, than of any other science. In contrast to the experimental sciences, mathematics does not contain any empirical terms or experimental methods for verifying its propositions. These propositions must be proved, i.e., logically deduced from a small number of axioms, accepted without proof. All the hereinmentioned specificities of mathematical knowledge especially clearly came to the fore with the transition in mathematics from the study of the quantitative relations among magnitudes and figures to the investigations into structures of the most diverse kind, which often have only a distant similarity with the traditional objects of classical mathematics. In connection with this, the most widespread notions about the nature of mathematical knowledge and the relation of mathematical objects and structures with the real world, were subjected to criticism.

The empirical notions about mathematics—according to which mathematical knowledge is essentially identical with the natural-scientific knowledge—were the first to be criticised. Moreover, the empiricists misinterpreted natural-scientific knowledge itself. For instance, the followers of classical empiricism viewed its theory to be inductive generalisation of experience. The defenders of similar inductive-empirical notions wanted, according to N. Bourbaki, "to compel mathematics to arise from the experimental truths". Evidently, the new stage of development of mathematics refutes these notions. However, N. Bourbaki so strongly stress the dominant role of rational-theoretic thought in this process, that they entirely forget about the objective source of emergence of mathematical ideas and theories. They do agree that, "of course, on cannot deny that most of these forms had a completely determinate inductive content when they emerged; but once they were consciously deprived of this content, it was possible to give them all their efficacy, which constitutes their strength, and it was made possible for them to acquire new interpretations and to fulfil their own role in data processing" [5, s. 259]. Clearly, all this is true, but from this it at all does not follow that experimental reality goes into the mathematical structures as a result of some predetermination or pre-established harmony.

Exactly in the same way, these structures should neither be considered to be *a priori* constructions of the human mind, nor conventions or agreements devised for ordering empirical data or for "economy of thought" *a la* the subjective-idealist philosophy of E. Mach. One cannot deny the existence of elements of convention, agreement and even of quite understandable economy of thought in these structures, in so far as any abstraction, as pointed out by Engels, is a shortening, and thereby it rids us of a mass of details. However, these elements do not play a self-sufficient role and can be correctly understood only in that case, when a structure is considered in the process of its historical emergence and development, wherein the empirical and the theoretical, the contentful and the formal and, the concrete and the abstract factors dialectically interact with each other.

When we deal with the ready-made mathematical structures, then at the first glance they do indeed appear to be *a priori* forms of knowledge, which turn out to be applicable to the study of very different contents. However, why does it so happen? Even the school of N. Bourbaki does not deny the fact that the initial concepts and structures of mathematics do have a fully determinate intuitive content; and this is evident from the aforementioned quotation. Many western philosophers of science stress the priority of form over content in all kinds of ways and that is why they view the mathematical structures as pure forms. One has only to attentively follow the genesis of these forms historically and logically, for the said illusion to vanish.

In fact, isn't there any connection and continuity between the primary structures of mathematics, which have an entirely determinate intuitive content, say the three dimensional space of Euclid, and the many dimensional or even infinite-dimensional abstract spaces? Don't these structures emerge thanks to the singling out of the deeper and more important properties and relations of the mathematical objects and structures under investigation? Considering the point of a many-dimensional space to be a vector, for which the ordered sequence of real numbers serve as co-ordinates, we postulate that for them the basic laws of operations over vectors, hold good. Owing to this it becomes possible to establish the connection and to differentiate between the three-dimensional and the many-dimensional spaces: all the laws of operations involving the vectors of the ordinary three-dimensional space do not hold good in the many-dimensional space. No less important is the fact that owing to this, the continuity in the development of mathematical knowledge, and the possibility of testing the conclusions of the many-dimensional space with the help of those of the three-dimensional, are established, since in the limiting case such conclusions must correspond to the results of ordinary geometry.

The situation is quite analogous with any abstract structure in general. The deepest properties and relations of the abstract mathematical objects are formulated in the axioms of these structures. It is, namely, that is why that these structures turn out to be applicable to the investigations of the most diverse phenomena and processes of the real world, and not only of those which served the emergence of the primary structures.

Having begun the work with such ready-made structures, a mathematician usually forgets those intuitive prototypes, which provided the impetus for the formation of the entire chain of successive abstractions and generalizations. Of course he is not bound to remember them, since that would complicate his work. Having proposed an abstract structure, he deduces the logical conclusions from the axioms, seeks various interpretations for them, establishes its connections with the other structures etc. In other words, the modes of application and elaboration of the mathematical structures are directly opposite to the historical path of their formation. Historically speaking, mathematical knowledge went from the separate, concrete system of objects, or interpretations, towards revealing their general structures, i.e. from the particular and the separate to the general, whereas in the process of further investigations it moves from the ready-made abstract structures towards the observation of various other interpretations.

Marx termed this process of development of mathematical knowledge, wherein its final point is taken as the beginning of the further movement of thought, the "inversion of method". He has illustrated it in detail in the "Mathematical Manuscripts", in the light of the emergence and use of the basic concepts of differential calculus. In the article entitled "On the Differential", he has shown, how, in course of the elaboration of the calculus, the symbolic differential co-efficient becomes an independent point of departure, but "with this, the differential calculus too appears as a specific kind of calculus, already operating independently upon its own ground...." [2, s. 55-57].

These ideas of Karl Marx about the "inversion of method" in the course of mathematical cognition, provides us with an opportunity to proceed correctly towards the solution of the problem of objective contents of the abstract structures of modern mathematics. At the first glance, their emergence appears to be *a priori*, but in reality it is the product of a protracted development, in course of which, the deepest and the most important properties of the structures are gradually brought to light. However, as soon as this cycle of development comes to an end and leads to the formation of the corresponding mathematical structure, this final point becomes the beginning of a new stage of mathematical cognition, connected with the elaboration of the theory of the given structure and with its application in the other sciences.

In so far as the entire historical process of emergence of the structures, usually goes on outside the field of vision of the modern mathematician, it is easy for him to have an illusion about the *a priori* character of the abstract structures or about some pre-established harmony among them and the empirical reality.

Roots of the idealist notions about the nature of mathematical cognition consist in this that, therein those connections between the abstractions and reality are ignored, which are realized in the process of application of mathematics in the natural sciences, technology and in the social-human sciences. Here, the contra-positing of "pure" mathematics and its application, is also to be largely blamed. This has promoted the cultivation of the idealist notions to the effect that, it is not the case that the abstractions and structures of mathematics are in agreement with the real world, but, conversely, it is the world and its regularities that come to conform to these abstractions and structures of mathematics. Criticising similar idealistic views about mathematics F. Engels pointed out that here, "as in every department of thought, at a certain stage of development the laws which were abstracted from the real world, become divorced from the real world, and are set up against it as something independent, as laws coming from outside, to which the world has to conform" [1, s. 38].

Another widespread point of view on the mathematical structures is connected with their conventionalist treatment. The famous French mathematician H. Poincaré is the founder of this approach; under the influence of the discovery of the new, non-Euclidean geometries he began to think that the axioms of geometry are conventional agreements and, while choosing them the mathematician is guided exclusively by the demands of convenience. We have seen that the abstract structures are defined by their axioms and, to that extent, here the mathematician enjoys considerable freedom, of choice, and of mutual combinations of the axioms and, that is why, in these structures the conventional moment is clearer than in the ordinary geometrical systems.

Understandably, one cannot deny the fact, that in the formation of the abstract structures, as in that of any other mathematical concept, the elements of choice and agreement do have a place. Without such choice, mathematical creativity would become meaningless, but freedom of choice does not signify a rule of arbitrariness. It is confined within the framework of necessity and, in mathematics in particular—constrained by the demands of logical non-contradictority of the axioms of the structure. But how can we be sure of their non-contradictority? In the light of the example of the geometry of Lobachevsky we have seen that the non-contradictoriness of its aximos can be proved by constructing its model with the help of the geometry of Euclid. In its turn, the non-contradictoriness of the geometry of Euclid can be proved with the help of an arithmetical model.

This process of proving the relative non-contradictoriness of the more abstract and new theories with the help of the old theories, with which we are more accustomed, is very characteristic of mathematical cognition. It is testified, firstly, by the fact that there exists a continuity and close inter-connection among the new and the old mathematical theories. Secondly, the freedom in the process of creation of the new mathematical structures, constrained by the demands of logical non-contradictoriness of the system of axioms, in essence signifies this, that the conventional elements play a subordinate role in mathematical cognition. The mathematician may substitute some axioms by others or seek out more general premises for his conclusions, but ultimately the correctness of his results are controlled by logic and by such well substantiated and corroborated theories, as the elementary geometry of Euclid and arithmetic, the truth of which have been tested in the centuries old practice of mankind.

Literature

1. *Marks K.*, *Engels F.*, Soch. 2-e izd., T.20 (Collected Works. Eng. ed. v. 25).
2. *Marks K.*, Matematicheskie Rukopisi. M., 1968.
3. *Lenin V.I.*, Poln. sobr. soch. T. 29 (Collected Complete Works, Eng. ed.. vol. 38).
4. *Bukur I.*, *Deleanu A.*, Vvedenic v teoriu kategorii i funktorov (Introduction to the Theory of Categories and Functors). M., 1972.

5. *Burbaki N.* (*Bourbaki N.*), Ocherki po istorii matematiki (Essays on the History of Mathematics). M., 1963.
6. *Dyson F.J.*, Matematika v fizicheskikh naukakh (Mathematics in the Physical Sciences) // Matematika v sovremennom mire (Mathematics in the Modern World). M., 1967.
7. *Foss A.*, Sushnost matematiki (Essence of Mathematics). M., L., 1923.

Source: Zakonomernosti razvitiya sovremennoi matematiki. "Nauka", M., 1987, s. 155-169.
About the Author: Georgi Ivanovich Ruzavin (1922-2012), Soviet philosopher and mathematician. Areas of specialization: Philosophy of Mathematics and Logic of Probability.
Other works:

1. O kharaktere matematicheskoi abstraktsii, 1961;
2. Osnovnye etapi razvitiya formailnoi logiki (soot. P.V. Tavantsova), 1962;
3. Induktsia i vcroyatnost. 1962;
4. Veroyatnostnaya logika i co roil v nauchnom issledovanii, 1964;
5. Induktsia i veroyatnost, 1965;
6. Induktivnye vyvody i veroyatnostnaya logika, 1967;
7. Semanticheskaya kontseptsia induktivnoi logiki. 1967;
8. O prirode matematicheskovo znaniya, 1967;
9. "Matematicheskie Rukopisi" K. Marksa i nekotorye problemy metodologii matematiki // *Voprosy Filosofii*, 1968. No. 12, s 59-70.

Reflections on seven themes of philosophy of mathematics

Vladimir Andreeivich Uspensky

PREFACE

An All-Union symposium was organised in the town of Obninsk, from the 26th to the 29th of September 1985, on the theme "The Regularities and Modern Tendencies of the Development of Mathematics". I took part in it. I was invited by Vladimir Invanovich Kuptsov. To him, to a large extent, goes the credit for the relaxed, creative and business-like atmosphere of this symposium. The papers were followed by intensive discussions, which continued in the so-called "round table" meetings. I did not read any paper, but look part in the discussions several times. Mikhail Ivanovich Panov thought that what I said was good enough for publication and, it is he who gave them the shape of a paper, to be included in a collection [of some of the papers read at the symposium] edited by him. It is thus that these "Seven Reflections" came into being. Here are the themes:

1. Is it true that in mathematics everything is defined and proved?
2. Is it possible to define the concept of natural number?
3. Is it possible to define the Series of Natural Numbers (written with capital letters)?
4. Is it possible to axiomatically define the concept of a series of natural numbers (written with small letters)?
5. Is it possible to prove, that Fermat's Great Theorem can neither be proved nor disproved?
6. What is a proof?
7. Can mathematics be made understandable?

1. Is it true that in mathematics everything is defined and proved?

Mathematicians are, as a rule, proud of the fact that they are mathematicians. For them, the source of their pride lies in their discipline—not so much in the usefulness of mathematics, as in the fact, that it is an unique field of knowledge, not resembling any other. Even the non-mathematicians are in agreement about its exclusiveness (thus not only the mathematicians themselves, but to their satisfaction, even those around them, recognise the greatness of the mathematicians). Indeed, it is considered to be generally acknowledged, that at least the three following traits belong to mathematics, and to it alone. Firstly, in mathematics, unlike in the other disciplines, all the concepts are strictly defined. Secondly, in mathematics—and again unlike in the other disciplines—everything is strictly proved from axioms. Thirdly, no other discipline has attained that level of respectful trepidation, at which mathematics remains not understood. Tutors of mathematics are hardly more in number, than those of all the other school-level subjects taken together, and yet there is nothing that one (of them) can really say about modern "higher" mathematics: it would be enough to open any monograph, or better still a paper in any journal. (Please note, that it often goes unnoticed, that the third trait indicated above clearly contradicts the first two).

When something becomes very well-known, then a suspicion creeps in: isn't this "something" a myth (well-known ideas do indeed possess an autonomous self-support mechanism). Let us attempt an, as far as possible, unbiased critical examination of the three, just indicated, well-known traits of mathematics.

We notice, firstly, that it is not possible to define all the concepts of mathematics. One is defined through the other, this other through a third etc.; we must stop at some place. (Mrs Prostakova rightly observed—"A tailor learned his trade from another, that one from a third, but from whom did the first tailor learn it?") A story goes, that once the famous mathematician from Odessa S.O. Shatunovsky was introducing ever newer concepts in a lecture and, while so doing he was repeatedly being asked: "And what is this and what is that?"; he lost his patience at long last, and asked in retort: "And what is 'what is'?".

Let us consider the structure of a defining dictionary in any language—Russian, English or any other. In it, one word is defined through another, that one through a third etc. However, since the words of a language are finite in number, emergence of a circularity is unavoidable (i.e., there emerges a situation, when, in the final count, a word is defined through and by itself). [Here it would be useful to think of a graph, in which the words are placed at the apexes and, when in the dictionary entry defining the word X one meets the word Y, then in that case an arrow goes from the apex X to the apex Y.] There is only one way of getting rid of such a circle: some words are to be left undefined. That is what is done in some of the dictionaries. [For instance, the words "thing" (in its principal meaning) and "all" have been left undefined in the defining dictionary of English language compiled by Hornby and Parnwell [8]. Unfortunately, such a dictionary has not yet been compiled for the Russian language.] Clearly, such is also the case with the concepts of mathematics. If, namely, one does not wish to permit a vicious circle, then one has to leave some of the concepts undefined. The question arises—how are these concepts going to be assimilated? Answer: through immediate observation, from experience, from intuition. There is no need to remind the reader, that the formation of general, abstract concepts in the human brain, is a complex process, which belongs more to the realm of psychology, than to that of logic. These concepts, which are assimilated not from verbal definitions, but rather from immediate personal experiences, are naturally called the *primary concepts* or categories of mathematics. The concepts of point, straight line, set, natural number etc. are examples of such categories.

One is certainly required to be careful while preparing a list of the categories (primary concepts) of mathematics (such a list can hardly be made fully precise). Otherwise the number of primary concepts will become unjustifiably large and, the principle of "Occam's razor" will be violated. Here let us consider, for example, the concept of a sphere. It is well-known that a sphere is the locus of points in space having a given fixed distance from a given point—which is the centre of the sphere. However, we can hardly find anyone, who came to know what is a sphere, first of all, from this definition. We must concede that a person assimilates the concept of a sphere in childhood—from the examples of a ball, a globe, a ball-bearing and a billiard ball. One learns the aforementioned definition of a sphere only in the classroom. There, the teacher does not always find the time to explain it to the learners that the sphere known to them since their early childhood and, the sphere about which they were taught at school—happens to be one and the same sphere. In consequence there grows the notion that: "everything is upside down in their physics and mathematics; perhaps even their sphere too goes upwards". The words just quoted were uttered by a "quite intelligent student", in justification of a statement made during a lesson, to the effect that a sphere put on an inclined plane starts rolling upward. This remarkable episode has been described in [10]. However, from what has been said above does it follow that since one comes to know about the concept of a sphere from experience, and not from a verbal formulation, the concept of a sphere should be considered a primary concept, one of the categories of mathematics?

It would appear, that the situation becomes more clear in the case of the more complex concepts of mathematics, which are further removed from experience, like for example, the concept of a group—surely, one would not regard the concept of a group to be a primary concept. However, the process of formation of the concept of a group in the brains of professional mathematicians is, perhaps, not very different from the process of formation of the concept of a sphere in the brains of people in general (which includes the mathematicians and the non-mathematicians): just as the concept of a sphere emerges as a result of numerous observations of various spheres, so did the concept of a group emerge as a result of examination of concrete groups—and only then was this concept fixated in a verbal formulation (obviously, at issue here is the emergence of the concept of a group in the collective experience of mathematicians, and not in the experience of an individual mathematician). That is why not the mode of emergence of a concept, but rather the mode of transmission of information about it within a system of knowledge, which should be considered to be the characteristic indicator of its primacy (categoricity). For elucidating what has been stated above, we shall imagine a situation where the carrier of a system of knowledge—in our case knowledge of mathematics—has to transmit his knowledge to others. Then he can tell others, what is a sphere or what is a group, by using the verbal definition of the corresponding concept. That is why these concepts are not categories. If, however, one is required to communicate, what is a set, or what is a straight line or what is a natural number, then that is done differently. For example, it is said that: all the chairs in this room constitute a set, and all the Ostriches beyond the Polar Circle constitute a set (academician P.S. Alexandrov's example), and all the irrational numbers within the interval [0,1] constitute a set; and later on, after having provided sufficient number of examples, it is said that: "these are all sets"—and thus there emerges the general concept of a set. Analogously: zero, one, two, three, four, five etc. are all natural numbers, and thus emerges the general concept of natural number. [It is high time for putting an end to the anachronism of beginning the series of natural numbers with one. In a pencil-box there are always some natural number of pencils—perhaps zero. A natural number is the cardinality (of the number of elements) of a finite set, in particular—that of an empty set.] (We see that while explaining the concept of natural number, there appears the word "etc."—implicitly or explicitly, and it could not have been otherwise in the case of primary concepts: first, a sufficient quantity of examples are indicated and, then we have the word—"etc.")

Thus, the first among the myths about mathematics—that, "in mathematics everything is defined"—collapses. Let us proceed to the second: "in mathematics everything is proved from axioms". In order to get convinced that such is not the case and, to thus blast this second myth too, it would be enough to open the classic textbook of school geometry by A.P. Kiseleyov, or any textbook of mathematical analysis for the technical colleges, or any university level textbook of the theory of numbers. In all these textbooks we come across theorems being proved, but hardly any axioms (save the axiom about the parallel lines—the fifth postulate of Euclid). The situation is somewhat enigmatic. Indeed, if there are no axioms, then on the basis of what are the proofs—say, of the theorems of the theory of numbers—put forward? Evidently, on the basis of common sense and some notions about the basic properties of the natural numbers; though these notions are identical for all persons, they have not been explicitly formulated in the form of a list of axioms. (How far they may be so formulated! Well, that is the theme of our next reflection).

It must be stated in all honesty that, in reality, in mathematics one pretty often comes across theorems, which are proved without basing the proofs upon any kind of axioms. The situation with the third trait of mathematics indicated by us—namely, with its non-understandability—happens to be more complex. It would be very easy to say that it is a myth; but if in respect of the first two traits it was enough to ask mathematics itself—one asks and gets a negative answer—then here, of course, it is pointless to turn to mathematics with the question: whether or not it is understood. A survey of social opinion indisputably situates mathematics at a prized spot in terms of the level of non-understandability. The reasons behind such an opinion happen to be the theme of a separate large-scale investigation. Any explanation of this phenomenon, it must be admitted, can only be that much objective, as much it is possible, in general, to be objective about the issues of social psychology. We shall not plunge into such a discourse here. In our last reflection we shall be making some remarks on this theme.

2. Is it possible to define the concept of natural number?

It can certainly be said that a natural number is the quantity of items in a finite totality. Evidently this formulation is compatible with the meaning (to be more precise, with one of the meanings) of the verb "to define" according to the "Defining Dictionary of the Russian Language" edited by D.N. Ushakov [5] ("to give a scientific, logical characterization, a formulation of any concept, to lay bare its (scientific) content"), as well as with the formulation found in the Philosophical Encyclopaedia [11] (the "definition" of an object—the results of the investigations about which are reflected in the corresponding concepts—"may be viewed as an (explicit and concise) formulation of the contents of these concepts"). Let us proceed now to the concepts behind the verb "to define" and the word "definition", from the position of a mathematician. In this case, we demand that a definition should contain exhaustive information about the concept defined—it must be so exhaustive as to permit a person having no earlier knowledge about a given concept, to form a correct understanding of it, solely on the basis of the definition provided. In that case can we assume, that a person who does not know anything at all as to what a natural number is all about (we are not talking here about the term, but, namely, about the concept), will be able to assimilate this concept from the first sentence of the present paragraph? That is very very doubtful: when one really does not know what is a number, then it is quite likely that he may take the words "quantity of items" to signify, say, their total weight and, the very concept of finite totality of items gets diffused when one considers very large totalities. Probably everyone would agree that trillion to the power trillion is a natural number, however, it remains a fact that this number is greater than the number of atoms in the whole Universe. It is not clear, as to how far appropriate it would be to talk about a finite totality of trillion to the power trillion quantity of items [16].

Thus, we shall be captiously demanding an exhaustive completeness from a definition, i.e. we shall be demanding that the concept being defined be expressed with the help of generally accepted syntactical constructions through other concepts, which serve as starting points for the definition under consideration. Taking into consideration what has been said above, let us attempt the following formulation: a natural number is the cardinality of a finite set. Three basic concepts are operative in this definition: 1) set, 2) cardinality and, 3) finite. The just mentioned formulation indeed appears to be the definition of a natural number, within the frame-work of those theories wherein these concepts are already somehow interpreted (in particular, declared to be uninterpretable or primary). Namely such a definition—such, in the ideational sense, right upto the inessential details—has been accepted in the "Éléments de mathématique" of Nicolas Bourbaki. (In this connection I would like to remind the reader that in Bourbaki's theory the full name of one [unity] requires tens of thousands of symbols, for the purpose of being written down) [6, p. 188]. Common sense, however, refuses to accept the concepts of set, cardinality and finite to be simpler than the concept of a natural number. Here we have the typical example of a definition of the simple through the complex.

One should not take the above-mentioned statements to be a criticism of N. Bourbaki and of the other authors, who put forward analogous formulations. Evidently, they, like other people, have some *a priori* notion of a natural number (apparently, *a priori* in respect of the definitions they propose, and not in respect of experience). They do not intend to give an explaining definition of the concept of natural number (i.e., a definition that could be used to teach a novice). Their aim is more modest and technical: to give a definition of this concept within the frame-work of an expounded axiomatic set theory. The concept of a function can be defined through the concept of a pair, and the concept of a pair can be defined through the concept of a function. It is clear that these intellectual constructions have hardly anything in common with the problem of explaining to the uninitiated, what is a pair and what is a function. The aim of the entire foregoing discussion is to lead the reader to the following almost self-evident idea. Let us set aside the mathematical and logical problematique, connected with the search for a definition of (it would be more correct to say—the attempts at representing, modelling) the natural number, within the framework of this or that axiomatic theory. Let us take up instead, the attempts at providing a "naive" explanation of the concept of a natural number—an explanation, that would enable one who does not know, to know what it is all about. Very soon we shall get convinced that such attempts are fruitless. We must admit that the natural number is a primary, (indefinable concept—it is one of the categories of mathematics.

3. Is it possible to define the Series of Natural Numbers (written with capital letters)?

Having failed in our attempts at defining the natural number (or, on the contrary, having succeeded in finding out that this concept is an undefinable category), let us now turn to the concept of the Series of Natural Numbers. When written with the big or capital letters, the Series of Natural Numbers is the totality of all the natural numbers. If we know what is a natural number and understand the words "totality of all", then we know what is the Series of Natural Numbers. Conversely, if we know the Series of Natural Numbers, then we can easily define a natural number as one of its elements. That is why the concept of the Series of Natural Numbers is as undefinable as the concept of natural number. (However, the sentence "The Series of Natural Numbers is the set of all natural numbers" may be viewed as a legitimate definition of the concept of the Series of Natural Numbers through the primary, undefinable concepts of a "natural number" and the "set of all".)

The reader would exclaim—"How come? And what about the axioms of Peano? Don't they define the Series of Natural Numbers?" Of course they don't, and what is more, if one understands the Series of Natural Numbers as we do—i.e., as the unique totality of some univocally understood essences called the natural numbers—then the axioms of Peano do not even pretend to do that. Indeed, let us see, how the axioms of Peano look: "Zero is a natural number, and zero is not the successor of any number, etc." Thus, these axioms are based on the concepts "zero" and "successor of" (here, immediate succession is at issue). However, they do not explain, and they cannot explain, what these concepts signify (i.e., what is "zero" and what is "successor of"), they only indicate the connections between these concepts. However, these axioms have been so formulated, that if the zero of these axioms is the ordinary Zero of the Series of Natural Numbers, and if the words "successor of" signify the immediate succession of one number after another in the Series of Natural Numbers (such that Zero is succeeded by Unity, and after Unity comes the figure Two etc.), then all these connections will be satisfied in the Series of Natural Numbers. [In order to stress the uniqueness, i.e., the absolute singularity of the terms of the Series of Natural Numbers—Zero. One (Unity), Two (the figure Two) etc.—we write them with the capital letters. The words "Zero", "One" (or "Unity"), "Two" (or the figure "Two") etc. are proper nouns in the absolute sense (just like the words "Sun", "Moon" and "Earth"), each of them has unique meaning—the quantity of elements of the empty, unit, two-element etc. sets. The "zero" axiom of Peano indicates a proper noun only relatively, within the limits of the given context, to be more precise—in the context of that structure, which has been described by these axioms. There are many such structures, and each one of them has its own zero.] In other words, Peano's axioms turn out to be true, correct statements upon their natural interpretation in the Series of Natural Numbers. However, evidently, they will be true, not only in the Series of Natural Numbers, but also in all the other structures isomorphic [on the concepts "isomorphism" and "isomorphic" we request the reader to go through the second of the two articles entitled "Isomorphism" in the 3rd edition of the Great Soviet Encyclopaedia [14]—V.A. Uspensky; readers of the present translation will find this article on "Isomorphism" at the end of the present theme 3.—Tr.] to the Series of Natural Numbers. For example, if the term "zero" in the axioms of Peano is interpreted as the smallest prime number, and the words "successor of"—as the [result of] a transition from one prime number to another immediately next to it, then under such an interpretation all the axioms of Peano turn out to be true. It appears that these axioms do not even permit us to distinguish the Series of Natural Numbers from the totality of all the prime numbers. I repeat, they do not pretend to do so. They claim to, it is said, "define the Series of Natural Numbers right upto isomorphism". To be more precise, the axioms of Peano define not one, but at once many mathematical structures, moreover they are all isomorphic to the Series of Natural Numbers and, consequently, isomorphic to each other. To be more precise, the axioms of Peano define the entire class of such structures. We shall call any such structure a series of natural numbers (written with small, or lowercase letters!). Thus, the Series of Natural Numbers is one of the carieses of natural numbers.

Briefly speaking, isomorphism of two mathematical structures is the mutually-univocal correspondence among the totalities of elements of the first and the second structure, retaining the operations and relations defined on these structures, In our example the isomorphism between the structure **N** (the Series of Natural Numbers with the operation "to follow") and the structure **P** (prime numbers with the operation "to follow") provides the following endless table:

0	1	2	3	4	5	6	...
2	3	5	7	11	13	17	...

In this correspondence the operation "to follow" is indeed retained: 6 follows 5, and simultaneously 17 follows 13, and in general in the upper row y follows x if and only if the corresponding terms of the lower row p_y and p_s (namely, in this order!) follow one after the other (follow in the sense defined for **P**).

It is sometimes said that the Series of Natural Numbers is the series

0, 1, 2, 3, 4, 5, 6, ..., 126, ...

but likewise it can be said that the Series of Natural Numbers is the series
zero, one, two, three, four, five, six ..., one hundred and twenty six, ...
or the series

$\overline{\underline{0}}$, I, II, III, IV, V, VI, ..., CXXVI, ...

[Isn't the persistent exclusion of zero from the series of natural numbers explained by the absence of the symbol $\overline{\underline{0}}$ in the traditional collection of symbols? Briefly speaking, aren't we situated at the level of the Latins on this question?]

Evidently, none of these serieses happens to be the Series of Natural Numbers (which consists of abstract quantitative categories and cannot be depicted), these are but the serieses of names designated for its terms, i.e. for the natural numbers. At the same time each of these serieses of names may be viewed as one of the serieses of natural numbers, written with small letters.

The situation with the Series of Natural Numbers is universal in character. For example, we have an analogous situation with the three dimensional Euclidian space, in which we live. Let us digress from the fact that most probably we live in non-Euclidian space, and generally speaking, we live not in the mathematical, but in the physical space—and these are different objects. [In this connection we must mention the fact that, most probably, the "physical" Series of Natural Numbers is something different from its mathematical model—the "mathematical" Series of Natural Numbers. On this issue see the deep-going but insufficiently appreciated essay [16] by P.K. Rashevsky.] Let us abstract from reality and imagine that we live in an entirely concrete three-dimensional Euclidian Space (we are again using capital letters, as we wish to stress the uniqueness of this space). However, it cannot be defined with the help of any number of axioms, it may only be "indicated with a finger". On the other hand, there are numerous systems of axioms (the most famous among them belongs to Hilbert) [3], defining this space "right upto isomorphism". The phrase within quotation marks indicates the fact that the given system of axioms defines an entire class of mutually isomorphic spaces, and that our "real" Euclidian Space happens to be one of them.

In general, no system of mathematical axioms can ever define any structure univocally, in the best of the cases they define it right upto isomorphism. (We speak here of "the best of the cases" as there are very important systems of axioms, which define, the class of non-isomorphic structures. For instance, the axioms of group theory define the mathematical structures called the groups, but all of them are not mutually isomorphic.)

Let us sum up. It is not possible to axiomatically define the Series of Natural Numbers. We may try to axiomatically define the concept of a series of natural numbers—i.e., the concept of any arbitrary structure, isomorphic to the Series of Natural Numbers. We shall be discussing these attempts in our next reflection.

ISOMORPHISM

It is one of the fundamental concepts of modern mathematics. It initially arose in algebra in connection with the algebraic systems, such as groups, rings and fields, but proved to be extremely significant for the understanding of the structure and domain of possible applications of every branch of mathematics.

The concept of isomorphism applies to systems of objects on which operations or relations are defined. As a simple example of two isomorphic systems consider the system $\boldsymbol{R}$ of all real numbers under the operation of addition $x = x_1, + x_2$ and the system $\boldsymbol{R}>0$ of positive real numbers under the operation of multiplication $y = y_1 y_2$. It turns out that the internal "lay out" of these two systems of numbers is identical. To show this we map the system $\boldsymbol{R}$ onto the system $\boldsymbol{R}>0$ by associating to the number $y = a^x$ $(a > 1)$ in $\boldsymbol{R}>0$, the number x in $\boldsymbol{R}$. Then the product $y = y_1 y_2$ of the numbers $y_1 = a^{x_1}$ and $y_2 = a^{x_2}$ which correspond to x_1 and x_2, will correspond to the sum $x = x_1 + x_2$. The inverse mapping of $\boldsymbol{R}>0$ onto $\boldsymbol{R}$ is given by $x = \log_a y$. From any proposition concerning the addition of numbers in the system $\boldsymbol{R}$ we can obtain a corresponding proposition concerning the multiplication of numbers in the system $\boldsymbol{R}>0$. For example, since in $\boldsymbol{R}$ the sum

$$S_n = x_1 + x_2 + \ldots + x_n$$

of the terms of an arithmetic progression is given by the formula

$$S_n = \frac{n(x_1 + x_n)}{2}.$$

it follows that in ***R*>0** the product

$$\boldsymbol{R}_n = y_1 y_2 \ldots y_n$$

of the terms of the corresponding geometric progression is given by the formula

$$\boldsymbol{R}_n = \sqrt{(y_1\ y_n)^n}$$

(raising to the n-th power in ***R*>0** corresponds to multiplication by n in ***R*** and extraction of the square root in ***R*>0** corresponds to division by two in ***R***).

As regards their properties, isomorphic systems are essentially the same. From an abstract mathematical standpoint, such systems are indistinguishable. Any system of objects S' that is isomorphic to the system S may be regarded as a "model" of S (modelling a system S by means of a system S'), and the study of the properties of S may be reduced to the study of the properties of the "model" S' of S.

The following is a general definition of the isomorphic system of objects such that each system has a number of relations and each relation involves a fixed number of objects. Let

$$F_k\ (x_1, x_2, \ldots), \quad k = 1, 2, \ldots, n$$

be the relations on S and let

$$F'_k\ (x'_1, x'_2, \ldots), \quad k = 1, 2, \ldots, n$$

be thc relations on S'. The systems S and S', with their respective relations, are said to be isomorphic if there exists a one-to-one correspondence

$$x' = \Phi(x) \quad x = \psi(x')$$

between the elements of S and S' such that

$$F_k\ (x_1, x_2, \ldots)$$

implies

$$F'_k\ (x'_1, x'_2, \ldots)$$

and vice versa. The correspondence is said to be an isomorphic map, or an isomorphism. [In the example cited above, the relation $F(x, x_1, x_2)$, where $x = x_1 + x_2$, is defined on the system ***R***, and the relation $F'(y, y_1, y_2)$ where $y = y_1\ y_3$. is defined on the system ***R*>0**, a one-to-one correspondence is given by the formulas $y = a^x$ and $x = \log_a y$.]

The concept of isomorphism arose in group theory where the fact that the study of the internal structure of two isomorphic systems of objects represents one rather than two problems was first understood.

The axioms of any mathematical theory determine the system of objects studied by the theory only upto isomorphism: a mathematical theory based on axioms, that is applicable to one system of objects is always fully applicable to another, Therefore, every axiomatic mathematical theory allows not one but many "interpretations" or "models".

The concept of isomorphism includes, as a particular case, the concept of homeomorphism, which plays a fundamental role in topology,

A particular case of an isomorphism is an automorphism, which is a one-to-one mapping

$$x' = \Phi(x) \quad x = \psi(x')$$

of a system of objects with given relations $F_k\ (x_1, x_2, ...)$ onto itself such that $F_k\ (x_1, x_2)$, implies $F_k'\ (x_1', x_2', ...)$ and vice versa. This concept also arose in group theory but later proved significant in most disparate branches of mathematics.

References

Kurosh, A.G., *Kurs Vysschei Algebry*, 3rd. ed., M-L., 1952.
Entsiklopedia elementarnoi matematiki, bk. 2, Ed. P.S. Alexandrov et al. M-L., 1951.
[*Source: The Great Soviet Encyclopedia.* A Translation of the Third Edition. Vol. 10, p. 465. Macmillan, 1976.]

4. Is it possible to axiomatically define the concept of a series of natural numbers (written with small letters)?

So then,we get down to the attempts at axiomatically defining the concept of a series of natural numbers, which is a structure isomorphic to the Series of Natural Numbers. As soon as we utter the word "isomorphism", already thereby it is proposed, that the relations and operations to be retained under this isomorphism have been indicated. Consequently, first of all we must indicate precisely, the relations and operations we wish to examine in the Series of Natural Numbers and in the serieses of natural numbers isomorphic to it. Among these operations we may include zero-place operations (i.e., individual constants; for example, the individual constant "zero" may be viewed as a zero-place operation) and one-place relations (i.e., properties). These earmarked operations and relations are to a significant extent arbitrarily indicated. For example, the Series of Natural Numbers (and thereby any series of natural numbers isomorphic to it) may be viewed: 1) as a structure only with the order relation "<", or 2) as a structure with an earmarked clement "zero" and the operation "transition to the next", or 3) as a structure, wherein, apart from the relations and operations already mentioned, the operations of addition and multiplication have also been earmarked.

For our purposes it would be most graphic not to indicate any operation, but only to stipulate the order relation "<". Thus we shall be viewing every series of natural numbers as a set, on which the binary order relation "<" has been defined. We shall be investigating, namely, the properties of such a mathematical structure.

Let us enumerate these properties. When the relation "<" is understood as an ordinary relation of order among natural numbers, then every property of the relation

“<” in an arbitrary series of natural numbers must (on the strength of the presence of isomorphism) hold good also in the usual Series of Natural Numbers. After this remark, let us now formulate some of these properties.

1. The relation “<” is transitive. Symbolically:

$$\forall x \, \forall y \, \forall z \, (x < y \wedge y < z \Rightarrow x < z)$$

2. The relation “< “ is anti-reflexive. Symbolically:

$$\forall x \, \neg \, (x < x).$$

3. The relation “< “ is symmetric. Symbolically:

$$\forall x \, \forall y \, (x < y \vee y < x).$$

The totality of these three properties simply affirms that “<” is a relation of strict linear order.

Before going ahead further, let us stop and think: strictly speaking, why are we listing these properties? Here’s why. We hope that having listed a number of properties, we shall be able to axiomatically define a series of natural numbers. In greater detail, our plan is as follows. At first we write out some of the properties characteristic of the Series of Natural Numbers. Then we declare these properties to be axioms and define a series of natural numbers as an arbitrary mathematical structure, satisfying the listed axioms. We do not exactly claim that a set defined with the given binary relation “<” satisfies our axioms (such a claim would be quite unrealistic), but we do claim that all such sets (with the given relation) turn out to be mutually isomorphic. In so far as the Series of Natural Numbers will satisfy our axioms (we shall be so chosing the axioms), the Series of Natural Numbers will be one of the pairwise isomorphic structures, satisfying these axioms, and this means that all these mutually isomorphic structures will also be isomorphic to the Series of Natural Numbers. If we succeed in attaining the goal just enunciated, then we should think that we have been able to axiomatically define a series of natural numbers.

Keeping in view the aim that we have put forward, can we remain satisfied with the three, properties-axioms listed out? Evidently, no. All linearly ordered sets satisfy these axioms, among them many are non-isomorphic and, consequently, wittingly non-isomorphic to the Series of Natural Numbers **N**. For example, the set of all real numbers **R** with the usual order relation will satisfy the three listed axioms. By comparing **N** and **R** we note that **N** has at least two such properties, which are absent in **R**.

These are:

4. **N** contains the smallest element. In symbols:

$$\exists x \, \forall y \, (x = y \vee x < y).$$

5. In **N** after every element x immediately follows some y. (“Immediately” means there is no third element between x and y.) In symbols:

$$\forall x\, \exists y\, (x < y \wedge \neg \exists z\, (x < z \wedge z < y)).$$

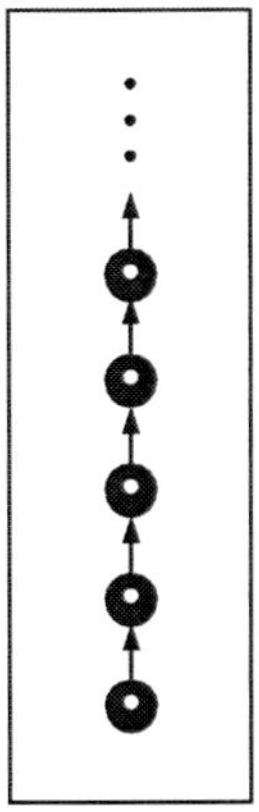

Fig. 1

These five axioms significantly narrow down the range of linearly ordered sets satisfying them. The Series of Natural Numbers, as well as the set of real numbers

$$0, \frac{1}{2}, \frac{2}{3}, \frac{3}{4}, \frac{5}{6}, \frac{6}{7}, \ldots \qquad (*)$$

(considered in the usual order), comply with these axioms. The existence of this, different from **N**, structure (*). satisfying the axioms 1-5, still does not constitute an hindrance to the view that these axioms provide an axiomatic definition of a series of natural numbers: since this structure is isomorphic to **N** (and, thus it can be identified as a series of natural numbers).

Figure 1 provides a graphic depiction of the order in (*) (and in **N**). However, it is easily noted that the structure (i.e., the set plus the order relation):

$$0, \frac{1}{2}, \frac{2}{3}, \frac{3}{4}, \frac{5}{6}, \frac{6}{7}, \ldots 10, 10\frac{1}{2}, 10\frac{2}{3}, 10\frac{3}{4} \ldots \qquad (**)$$

too satisfies the axioms 1-5. Figure 2 gives a graphic depiction of this ordered structure. In this structure two elements (0 and 10) do not have any immediate predecessors. Let us fixate this situation in the following axiom 6.

6. If two elements x_1 and x_2 do not have any immediate predecessor, then they are equal. In symbols:

$$\forall x_1, \forall x_2\, \{[\neg \exists y_1\, (y_1 < x_1 \wedge \neg \exists z_1\, (y_1 < z_1 \wedge z_1 < x_1))] \wedge$$
$$\wedge [\neg \exists y_2\, (y_2 < x_2 \wedge \neg \exists z_2\, (y_2 < z_2 \wedge z_2 < x_2))] \Rightarrow x_1 = x_2\}.$$

Axiom 6 eliminates the structure (**), but does not eliminate the structure

$$0, \frac{1}{2}, \frac{2}{3}, \frac{3}{4}, \ldots, \ldots$$

$$\ldots 9 + \frac{1}{m}, 9 + \frac{1}{m-1}, \ldots 9\frac{1}{4},$$

$$9\frac{1}{3}, 9\frac{1}{2}, 10, 10\frac{1}{2}, 10\frac{2}{3}, \ldots, 10 + \frac{n-1}{n}. \qquad (***)$$

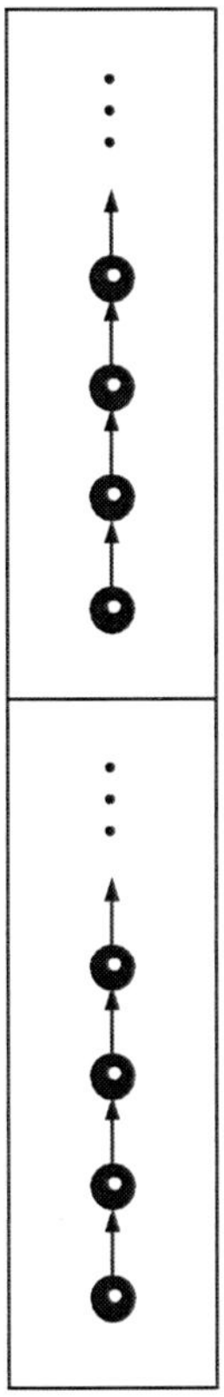

Fig. 2

It is evident that the structure (***) is not isomorphic to a series of natural numbers.

Like the horizon, our goal is moving further and further... It appears to be unattainable. The following noteworthy fact appears to be the case: however many axioms may we write out, using the logical symbols, the symbol "<" and the variables, covering the elements of the structure being defined—there will always be a model of the totality of listed axioms, non-isomorphic to a series of the natural numbers. In view of the fundamental importance of this fact (signifying the impossibility of axiomatically defining a series of natural numbers by using the means indicated), let us describe it in greater detail.

The following symbols belong to the alphabet of the formalized symbolic language, which we are using for listing the axioms:

1) the symbols of punctuation: the left hand side bracket "(" and the right hand side bracket")",
2) the logical symbols "$\rceil$", "$\wedge$", "$\vee$", "$\Rightarrow$", "$\forall$", "$\exists$", "$=$",
3) the individual variables $x, y, z, u, v, w, x_1, y_1, z_1, u_1, v_1, w_1, \ldots,$
4) the symbol "<".

Formulas are composed with the help of these letters, according to the natural, but easily formulated syntactical rules. Simplest examples of such formulas are:

$$x < y \vee y < x;$$
$$\forall x\ (x < x);$$
$$\exists x\ \exists y\ (y < x \Rightarrow y < < x);$$
$$\exists y\ (x < y);$$
$$\forall x\ \exists y\ (x < y).$$

Now, let us take any set, with any binary relation denoted by "<" defined on it (it is not obligatory for the relation to be one of strict order). We shall be calling all such sets with the relation "<", a *structure with the label* <. Thus, a structure with the label < consists of a set (called the *carrier of the structure*) and the relation "<". Let us fix a carrier of the structure as the domain of change for each individual variable. Then every formula becomes either a sentence, as the second, third and the fifth formulas of the just mentioned list are, or sentential forms, like the first and the fourth formulas of the same list. Those formulas which turn into sentences are called *closed*; we shall be considering only them in future. It is not difficult to notice, that the property of "being closed" in respect of a formula, does not depend on the structure, wherein we are examining the said formula; this properly can be defined purely syntactically, according to the external form of the formula.(Closure consists of this that all the variables must be bound by quantifiers.) It is said in respect of a (closed) formula—when considered in a given structure—which becomes a true sentence, that it is *true in the given structure* or that it is *satisfied in the given structure*; and about the structure it is said, that it is *satisfied by the given formula.*

The structure **N**—our usual Series of Natural Numbers with the usual order relation—may be singled out from among the structures with the label <, we shall call any closed formula, turning into a true sentence when interpreted in the structure

N—an axiom. Thus, however many—finite or infinite number of—axioms may we write out, there will always be such a structure with the label <, which, firstly, satisfies all the listed axioms, and is secondly, isomorphic to **N**.

Thus, it so turns out, that a series of natural numbers cannot be defined axiomatically: since to define **N** axiomatically is to list such a system of axioms, as would define **N** upto isomorphism (this, in its turn, means, that any two structures satisfying all the listed axioms, are isomorphic).

"But excuse me"—the reader will again object—"the axioms of Peano do define the Series of Natural Numbers upto isomorphism. Peano's system of axioms is categorical, and this signifies, that all the models of it are isomorphic". [Any structure satisfying each of the axioms of a system is called the *model* of that system or list of axioms.] A little patience! We shall look into the axioms of Peano too.

However, now we shall discuss another question. Not merely the order relation "<", but a numerous set of other relations and operations are defined on the Series of Natural Numbers. Among them there are the two-place (or binary) relation of divisibility of two numbers; the three-place (or ternary) relation "$x + y = z$"; the one-place (or singular) relation of "being a prime number" (let us recall that we treat properties as one-place relations), there we are using the etymologically more correct term "singular", following W.V.O. Quine, instead of the now widely used term "unary"; see: 7. Church, note 29); the two-place operation of addition; the two-place operation of multiplication; the two-place operation of involution ($0^0 = 1$); the one-place operation of following immediately (as is customary, we shall indicate it with the prime symbol, such that, $0' = 1$; $13' = 14$); the constants 0, 1, 2, 3, 4, ... (let us recall that we treat the constants as zero-place operations); the four-place operation $[\log_{u+1}(z! + y^{x \cdot z + u})]$ (here, as usual, through $[a]$ we indicate the integral part of the number a); and many others. We have adduced only a few examples, and altogether a countless number of operations and relations are defined on **N**. In order to define the concepts of a structure isomorphic to **N**, we must at first separate out some operations and relations from among them (theoretically it is possible to take into consideration all of them) and examine the isomorphism, namely, in respect of these isolated operations and relations. Indeed, that is why there does not exist the concept of a series of natural numbers as such, but only the concept of a series of natural numbers in respect of a given list of operations and relations. Earlier we have examined the concept of a series of natural numbers in respect of a list, wherein there were no operations at all, and there was but one relation—the relation of "lesser than".

In the context of our investigations, the operations and relations singled out in the set are called—*labelled*, and the list of such operations and relations—a label. To be more precise, not the list of these operations and relations themselves, but the list of their names, is called a label, (this distinction is very important in itself) but for our purposes it is not quite essential, and it would be easier for us not to notice it. A set with some singled out operations and relations, forming the list σ, is called the *(mathematical) structure with the label* σ. Now we can say that any series of natural numbers is a structure with this or that label σ. That is why, we should be speaking not about a series of natural numbers in general, but about a series of natural numbers with the label σ. So far we have examined the case, when

$$\sigma = \{<\}.$$

Perhaps the poverty of this label is the reason behind the failure of our attempt at axiomatically defining a series of natural numbers? Let us broaden the label and see what happens. First of all, let us add to "<", the constant "0" (for denoting the smallest element in respect of the order "<") and the prime symbol "′" to indicate the operation of immediate succession. In the Series of Natural Numbers **N**, these objects come under the following axioms (properties) 7 and 8 (compare the properties 4 and 5, which follow from the properties 7 and 8):

7. $\forall y\ (0 = y \vee 0 < y)$;
8. $\forall x\ (x < x' \wedge \urcorner \exists z\ (x < z \wedge z < x'))$.

Any series of natural numbers with the label $\{0, ', <\}$ is by definition isomorphic to **N**, since isomorphism is considered in respect of $\{0, ', <\}$. That is why any such series of natural numbers consists of the elements 0, $0'$, ..., ordered as follows:

$$0 < 0' < 0'' < 0''' < \ldots$$

Remarks. We must be aware of the fact that every series of natural numbers has its own 0, own ′ and own <, i.e., own element indicated by 0, own operation signified through "′" and, own relation denoted by "<". Strictly speaking, for every series of natural numbers we shall devise its own symbols for these objects—for example, if we are considering a series of natural numbers **M**, then it is necessary to add this letter "**M**" as an index to the symbols "0", "′" , and "<". This strictness provides some convenience. However, the absence of strictness also gives rise to some convenience. In the given case, the convenience from lack of strictness is considered to be greater, and that is why one and the same "0" is used to signify various elements (but in every series of natural numbers it denotes one and only one element; in particular, the cardinality of the empty set in the Series of Natural Numbers). Analogously for "′" and "<". These remarks are valid not only for a series of natural numbers, but also for any structure with the label $\{0, ', <\}$, not bindingly isomorphic to **N**.

Now we shall see, how an arbitrary structure with the label $\{0, ', <\}$, subordinated to the axioms 1-8, looks (axioms 4 and 5 follow from the axioms 7 and 8, but that is not a great calamity). Evidently, it is a linearly ordered set, wherein 0 is the least element, $0'$ is the element immediately following 0 (such that there is nothing between 0 and $0'$). $0''$ is the element that immediately follows $0'$ etc. All these elements—0, $0'$, $0''$, $0'''$—form the initial cut of our structure. This initial cut is called the *standard* part of the structure, and the remaining part (it may even be empty) is called *non-standard*. The standard part is isomorphic to **N**. Had there been nothing but this standard part in any structure with the label $\{0, ', <\}$, subordinate to the axioms 1-8, then we would have attained our goal: axioms 1-8 in their totality would have given us the axiomatic definition of a series of natural numbers we are looking for, to be more precise—they would have given us the axiomatic definition of a series of natural numbers having the label {0, ′, <}.

This, however, is not the case. The structure graphically depicted in figure 3—say, the one like (***), where

$$0' = \frac{1}{2}, \left(\frac{1}{2}\right)' = \frac{2}{3}, \left(9\frac{1}{4}\right)' = 9\frac{1}{3} \text{ etc.,}$$

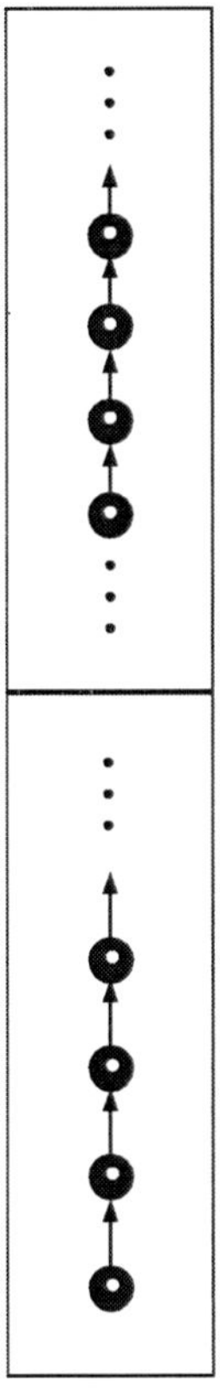

Fig. 3

satisfies the axioms 1-8, but it is not isomorphic to N: it contains a non-empty non-standard part (in figure 3 this non-standard part has been depicted in the upper rectangle, in (***) this non-standard part consists of the elements of the form $9+\frac{1}{m}$ and $10+\frac{n-1}{n}$). What is more, it turns out that no system of axioms can give us a series of natural numbers with the label {0, ′, <} , since the structure depicted in figure 3 will always be a model for such axioms.

Perhaps it is still a case of poverty of the label? What will happen if we add addition and multiplication and consider a series of natural numbers not having the label {0, ′, <}, but one with the label {0, ′, <, +, ·)?

Would it be possible to make a list of axioms for such a richer label, which would define the concept of a series of natural numbers having this label—i.e., from *all* the structures with this label, would it be possible to single out those structures which are isomorphic to **N** in respect of 0, ′, <, +, · ? It appears that no, it cannot be done. Whatever be the totality of axioms—finite or infinite—made out by us, for this totality there will always exist a structure [with the label {0, ′, <, +, · }], non-isomorphic to **N**. [When we speak of axioms, we have in view a symbolic language, like the one described above for the label {<}; only now, together with "<", the alphabet contains "0", "′", "+" and "·".]

What is more, whatever be the label chosen and whatever be the system of axioms chosen for this label, there will always exist a model of this system of axioms, not isomorphic to **N**. Such, non-isomorphic to **N**, models are called non-standard, and the axioms listing the properties of a series of natural numbers (especially, when + and · enter into the label), are called the *axioms of arithmetic*. That is why, we may re-state what has been stated above as follows: there exists a non-standard model for any system of axioms of arithmetic.

If the axioms 1-8 or something equivalent to them enter into our axioms, then it is possible to single out the standard part 0, 0′, 0″, ... in any model; in this case, non-standardness of a model signifies the non-emptiness of the non-standard part. This non-standard part may turn out be more complexly constructed, than the case depicted in figure 3. From the standpoint of order, the non-standard part depicted

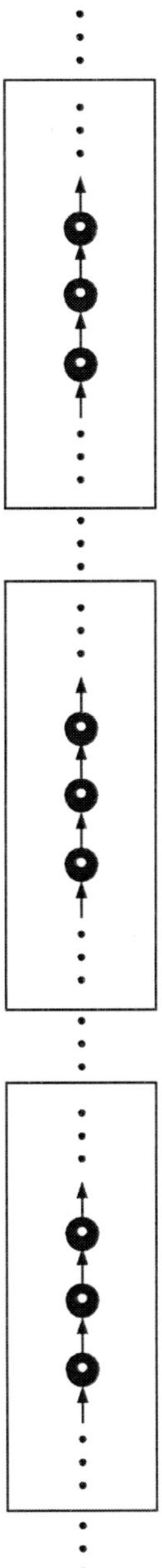

Fig. 4

in figure 3 is similar to the set **Z** of all whole numbers. In the case of natural axioms for a label that includes the operation of addition, the non-standard part of any denumerable structure (i.e., of one containing denumerable number of elements), satisfying these axioms, assumes a look, which we have (not very successfully) attempted to depict in figure 4. In this diagram we have attempted to somehow depict the following idea: a very large (denumerably infinite) number of examples of the sets of whole numbers **Z** are taken up, and these examples arc arranged like the set of all rational numbers **Q**.

Thus, it is not possible to produce a system of axioms, defining the concept of a series of natural numbers (with any label whatsoever). To our knowledge, a more detailed interpretation of this statement is as follows: whatever operations and relations defined on **N**, may be chosen, there cannot be such a system of axioms, all models of which are isomorphic to **N**, in respect of these operations and relations.

Now we shall answer the question: "But what about the axioms of Peano?".

With inessential changes, the classical axioms of Peano are us under. Here, the label $\{0, '\}$ is being considered .Three axioms have been formulated:

I. $\neg \exists x \, (x' = 0)$;

II. $\forall x \, \forall y \, (x' = y' \Rightarrow x = y)$;

III. The Axiom of Induction.

We have, till now, only named the third axiom of induction, but have not described it. Now we describe it:

$$\forall P \, \{[P(0) \wedge \forall x \, (P(x) \Rightarrow P(x'))] \Rightarrow \forall x \, P(x)\}.$$

When we look at the axiom of induction we notice that together with the usual individual variable it contains yet another variable P, we shall explain the meaning of this variable. First of all, let us recall that the semantics of a formula (i.e., the meaning attached to it) emerges only after the mathematical structure corresponding to the label is produced. In particular, in order to find out the meaning of the axioms of Peano (of the formulas I-III), we must produce some structure with the label $\{0, '\}$ i.e., a set with a singled out clement, indicated by "0" and a singled out one-place operation, indicated by "$'$". Then the domain of change of the variable x is at once defined (like that of any individual variable)—it is the set of all elements of the structure under consideration. What is the domain of change of the variable P like?

The variable P is of a special type, we have not met with the like of it hitherto in our enunciation. Its domain of change consists of all the possible properties

(= one-place relations), defined on the structure under consideration, i.e., the properties of the elements of this structure.

The concept of property is a primary concept; it is grasped from examples. The property of being even is defined on the natural numbers—every number may be either even, or odd. It is inessential that there are even as well as odd numbers; we may construct a situation, where all the numbers are even; what is important is that it should make sense to ask in respect of every number, whether it is even or odd. The property of being green is not defined on a series of natural numbers; for a number the talk of "being green" is pointless. The Series of Natural Numbers possesses the properties formulated above, as a whole. Relations, too, may possess properties: example—the relation of transitivity. But at the given moment we are interested only in the properties of the elements of the structure under consideration (for which the axioms of Peano are satisfied). It is these properties, namely, that can provide the values of the variable P.

The fact that an element a possesses the property Q is described as Q(a). If a property Q is defined on the elements of some set M, then it is possible to introduce for consideration a sub-set of this set, K—consisting of those and only those elements of M, which possess the property Q

$$x \in \mathrm{K} \Leftrightarrow \mathrm{Q}(x). \qquad (+)$$

Conversely, for every sub-set K it is possible to introduce the property Q: of "being an element of K", and again the correspondence (+) will be satisfied.

Thus, a property and a sub-set are almost the same: "the language of properties" and "the language of sub-sets" are trivially inter-translatable. For example, the axiom of induction would look as follows in the language of sub-sets:

$$\forall P \{[0 \in P \wedge \forall x (x \in P \Rightarrow x' \in P)] \Rightarrow \forall x (x \in P)\}.$$

Thus, in the axiom of induction, the domain of change of P is the totality of all the properties defined on the structure under consideration. We shall now see, how this axiom is utilised to ascertain the fact that a structure satisfying the axioms of Peano, is isomorphic to **N**. Thus, let a structure with the label {0, ′} satisfy the axioms I-III. Axioms I-II ensure the presence of a standard part $\{0, 0', 0'', 0''' ...\}$ in the structure. Now let us apply the axiom of induction, having taken as a value of the variable P, the property P_0 of the elements of the structure: "to belong to the standard part". The axiom says that something is true for all P_0, in particular for this P_0. Thus, it occurs that

$$[P_0 (0) \wedge \forall x (P_0, (x') \Rightarrow P_0(x')] \Rightarrow \forall x\, P_0(x).$$

The premise enclosed within square brackets is evidently true (0 belongs to the standard part, and if x belongs to the standard part, then x' too belongs to it); that is why $\forall x\, P_0 (x)$, i.e., all x (all elements of the structure!) belong to the standard part. We have already noted that the standard part is isomorphic to **N**. This concludes the proof of this that the structure under consideration is isomorphic to **N**.

Thus, any structure satisfying the axioms of Peano, is isomorphic to **N**, and consequently, these axioms define the concept of a series of natural numbers of the label $\{0, '\}$. Apparently, this situation contradicts our repeated announcement to the effect that it is not possible to formulate a system of axioms with such properties.

There are no contradictions here, and here's the reason why. Earlier we were speaking only about the properties of the Series of Natural Numbers, which could be expressed through a definite linguistic means—in other words, we were talking about some axioms written in a definite language. This language contained only one type of variables—the individual variables x, y, z, ... The essence of these individual variables lies in the fact that upon being interpreted on any structure, each one of these variables gets one and the same set as its domain of change—it is the set of all elements of the structure under consideration. Another kind of variable—the variable P—takes part in the axiom of induction. Its values are not the elements of the structure under consideration, but a property of these elements (in other words—the one-place predicates defined on these elements, whence the variable P itself is called a *predicate*, to be more precise—*a predicate variable of valency* 1). Thus, the axiom of induction is a formula of another, extended language; this language is more extended than the narrow language so far considered. (Narrow because it contains only individual variables). When we said that there is no system of axioms, which fully characterizes a series of natural numbers, then we had this earlier, narrow language in view.

Of course, an explanation has been provided, but it hardly satisfies any one. What if it is not possible to write out a system of axioms for a series of natural numbers, in some language? It is, as they say, "not a fact from the biography of a series of natural numbers, but rather one from the biography of that language". Simply put, a narrow language is bad, and look, now we have found a good, extended language, in which it is possible to write the adequate axioms for a series of natural numbers.

Everything, however, is not that simple. Crudely speaking, the situation is just the opposite: a narrow language is "good", an extended one—"bad".

Let us attempt an explanation of the situation. We shall begin with terminology.

The formulas, wherein all the variables are individual are called *elementary* formulas, and the language that permits of only the elementary formulas, is called an *elementary* language. In the given context, the synonym for the term "elementary" is the term "1st order" or "first order". All the axioms considered above, save the axiom of induction (i.e., the axioms 1-8 and I-II) were elementary axioms, i.e., elementary formulas. There exists no (neither finite, nor infinite, and besides of any label) system of elementary axioms, which would satisfy the Series of Natural Numbers **N** and, all the models of which would be isomorphic to **N**.

There are non-elementary formulas, but they belong to a non-elementary language. Variables of a more complex nature are permitted in this language—predicate variables of valency 1, properties (= one-place relations) serve as their values; predicate variables of valency 2, binary (= two-place) relations serve as their values etc., and also, functional variables (any one-place operation, like, say, "to follow", may serve as the value of a functional variable of valency 1, and any two-place operation, like, say, addition, may serve as the value of a functional variable of valency 2). The axiom of induction is an example of a non-elementary formula. A more precise non-elementary language, having the possibilities just described, is called a *2nd order language*: this means, that it admits of variables covering relations and operations (what sort of relations and operations, that must be defined on the elements of the structure), but does not consider more complex variables, as the values of which may serve, say, the properties of operations or operations on relations (or the properties of relations such as "transitivity"). The axiom of induction serves as an example of non-elementary formula of a 2nd order language (or simply, as an example of a 2nd order formula).

A second order language is the simplest of all non-elementary languages.

One would think—and the presence of the axioms of Peano somehow confirms this—that it is possible to have a system of non-elementary axioms of 2nd order (i.e., axioms, written in the form of formulas of this non-elementary language), defining the concept of a series of natural numbers, in the following precise sense:

1) **N** is a model of this system; and
2) any model of this system is isomorphic to **N**.

However, here arises an unexpected, but quite fundamental semantic (one may even say, epistemological) difficulty. The fact is this, that already in the case of a 2nd order language (not to speak of the more complex non-elementary languages), the very concept of a model loses its essential clarity. The following example, connected with the so-called problem of the continuum, illustrates this situation.

It is well known that the quantity of elements of any set is called the *cardinal number* or the *cardinality* of that set. The concept of a cardinal number or cardinality is a generalisation of the concept of natural numbers, in so far as the natural numbers are the cardinalities of finite sets. From among the infinite cardinalities the following two are singled out: the cardinality of the set of all natural numbers and the cardinality of the set of all real numbers. The first is indicated by $\aleph_0$ (read "aleph-null") and is called the denumerably-infinite (or infinitely denumerable) cardinality; the second is indicated by $\mathbf{c}$ (small gothic "c") and is called the cardinality of the continuum. Evidently, $\aleph_0 < \mathbf{c}$. The famous *continuum problem* consists of explaining whether or not there exists any intermediate cardinality, i.e., a cardinality satisfying the inequality

$$\aleph_0 < m < \mathbf{c}.$$

The famous continuum hypothesis consists of this, that such a cardinality does not exist. (On the strength of the results of K. Gödel and P. Cohen) it is well-known that it is neither possible to prove, nor disprove the continuum hypothesis. While speaking of "proving" and "disproving", we have in view all the conceivable means permitted in modern mathematics. Thereby, the question of the very meaning of the continuum hypothesis remains unsettled. Indeed, the meaning of such a statement is taken to be vague—its truth or falsity cannot be determined in any way. This extraordinary situation is radically different from such situations, often to be met with, when we simply do not know something (though we understand the question very well). [The security of clarity in understanding a question lies in the clarity of understanding the possible answers.]

It appears that we may write out a formula of the 2nd order, which then and only then has a model (i.e., such a structure, for which it becomes true), when the continuum hypothesis is true.

It is also possible to write such a formula of the 2nd order, the existence of a model of which is equivalent, on the contrary, to the existence of an intermediate cardinality, i.e., to the truth of negation of the continuum hypothesis. [*A note for the specialists*. Examples of such formulas, known to the present author, contain predicate symbols of valency 2. However, if in the axiom of induction we change the prime symbol "′" into a predicate symbol, then this axiom too would contain a predicate symbol of valency 2.] Thus, for the formulas of 2nd order, the question of existence of their models may turn out to be as hazy as the continuum hypothesis itself.

That semantically so vague a language would be able to serve as a satisfactory means for axiomatically defining something—in particular, a series of natural numbers—appears to be doubtful.

Indeed, if we analyse the use of the axiom of induction in the process of proving that any model of the axioms I—III is isomorphic to **N**, then we shall see, that here, in essence, we are using that very concept of a natural number, which we still only intend to define axiomatically. Our property P_0 signifies "to have the form 0″... ′ ". The dots in the expression "0″ ...′ " are only a substitute for the general notion of a natural number. And it is not possible to depict the property P_0 without an *a priori* notion of the natural numbers or without substituting that notion by dots or by the expression "etc.".

5. Is it possible to prove that Fermat's Great Theorem can neither be proved nor disproved?

We have mentioned the continuum problem at the end of our last reflection. It is one of the major problems agitating the intellect of mathematicians. In the famous report entitled "Mathematical Problems", read by the great Hilbert in the year 1902, at the International Congress of Mathematicians in Paris, it was mentioned as the first problem. We have already noted that the continuum problem turned out to be unsolvable: it is neither possible to prove, nor disprove the continuum hypothesis.

While enumerating 23 basic problems of mathematics, Hilbert did not mention the problem of proving (or disproving) the Great Theorem of Fermat. Evidently, Hilbert did not consider it to be important enough. At the same time there is no doubt that it is the most famous among the unsolved problems of mathematics. And besides, unfortunately, it is unique among the unsolved problems, known to the wide mass of non-mathematicians. We wrote "unfortunately", since professional mathematicians spend an appreciable percentage of their time, studying and refuting the essays of the Fermatists—the name given to people, who do not have the necessary mathematical preparation, but who think that it is, namely, they who have proved Fermat's theorem.

Strictly speaking, Fermat's theorem should not be called a theorem. The "Matematicheskaya Entsiklopedia" [22] defines a theorem as a "mathematical statement, the truth of which has been established through proof".

A proof has not yet been found for Fermat's "theorem". [However, not every one supports this point of view. Thus, Viktoly Invanovich Budkin states in p. 45 of his book "Methodology of Cognition of "Truth". A proof of Fermat's Great Theorem" (Yaroslavl: Upper Volga Publications, 1975. pp. 48): "13 generations have passed, and yet Fermat's Great Theorem still remains unproved. Only in the present work, a complete proof of the theorem is being given in its general form".]

What is more, the same "Matematicheskaya Entsiklopedia" contains an article entitled "Fermat's Theorem", in its 5th volume (the same volume that contains the aforementioned definition of a theorem). We too shall be using this generally accepted but imprecise term—though we admit that it would have been more correct to speak of the hypothesis of Fermat.

Many factors contributed to the popularity of Fermat's theorem among the non-professionals. These include: 1) the authority of its author: it has been stated by one of the originators of the theory of numbers—the famous French mathematician Pierre de Fermat; 2) the respect due to age: it was stated at around 1630; 3) the romantic circumstances of its formulation: Fermat wrote it down on the margins of the 1621 edition of the "Arithmetic" of Diophantus [the eighth problem of the second book of the "Arithmetic" of Diophantus reads—"To decompose a given square into two squares"; Fermat made the following comment on this problem—"On the contrary, it is not possible to decompose any cube into two cubes, any biquadratic into two biquadratics, and in general no power greater than the square, into two powers having the same index. I have discovered a truly wonderful proof of this, but these margins are too narrow for it"; this proof could not be found among the papers of Fermat]; 4) the setting up of a prize of 100,000 German marks for providing a proof of Fermat's theorem, in 1908 by Wolfskehl (naturally, the "pleasant" fact of the institution of a big prize became much more well-known, than the "unpleasant" fact of its complete depreciation as a result of the post-first-world-war inflation); and 5) the simplicity of its formulation.

Of course, the first four factors could not have been effective in tandem, had not the formulation of Fermat's theorem been so popular. It is as follows: Whatever be the integer n, greater than 2, the equation

$$x^n + y^n = z^n$$

has no positive integral solution.

We see, that the equation present in the formulation of Fermat's theorem may be viewed as an equation with three unknowns—x, y and z. In sofar as n may assume any of the values 3, 4, 5, 6 etc., here, in fact, we have an infinite series of equations, and it has been stated that none of them has solutions in such integral x, y and z that $x > 0$, $y > 0$ and $z > 0$. From the point of view of logic it would be more natural to consider the equation

$$x^n + y^n = z^n$$

as an equation with four unknowns x, y, z, and n. Then, Fermat's theorem would state that this equation has no integral solutions for $n > 2$, $x > 0$, $y > 0$ and $z > 0$.

The search continues for the proof(s) of Fermat's theorem*. Theoretically speaking, the search for its refutation could also have continued, but that is not happening. The situation with the hypothesis called "Fermat's theorem", is significantly different from the situation in respect of the continuum hypothesis: we know, that for the continuum hypothesis it has been proved that it can neither be proved, nor disproved (to be more precise, in 1939 Gödel showed that it cannot be disproved , and in 1963 Cohen showed that it cannot be proved). For the hypothesis (theorem) of Fermat, such a proof—the proof that it is neither possible to prove, nor disprove it—does not exist. The question arises: whether this proof does not exist so far (with the hope that it will be obtained in future) or is it in principle impossible? Had this proof been obtained, it would, undoubtedly, have been of great use for mathematics, as it would have closed, once and for all, the floodgates in the face of the flow of ignorant attempts to prove the theorem of Fermat. Unfortunately, such a proof is not possible. It is true, that there remains a theoretical possibility of proving that Fermat's theorem cannot he proved. The appearance of such a proof too would have closed the aforementioned floodgates—but then, perhaps, there would have emerged a flow of attempts to disprove Fermat's theorem (for example, by way of producing, in an oblique manner, four astronomically large numbers n, x, y, z for which the required equation would be practically unverifiable).

Thus, we are assuming, that (a) there exists a proof to the effect, that Fermat's theorem cannot be proved; (b) there exists a proof to the effect, that Fermat's theorem cannot be disproved.

Now, our aim is to show, that (a) and (b) are incompatible, i.e., it is not possible for these two statements to be true at the same time. In fact we find that (b) is incompatible even with the weaker-than-(a)-statement (a_1): "Fermat's theorem cannot be proved". We shall show, namely, that from (b) the existence of a proof of Fermat's theorem follows and thereby (a_1) is negated.

First, some preliminary remarks. Let us agree to call, any four natural numbers n, x, y, z such that $n > 2$, $x > 0$, $y > 0$, $z > 0$ and $x^n + y^n = z^n$, the *Fermat four*. Fermat's theorem states that the *Fermat four* do not exist. Disproving any theorem is to prove its negation. Disproving Fermat's theorem would mean proving that the *Fermat four* exist. [As before we are using inexact terms and identifying the word "theorem" with the word "statement" and not with the expression "proved statement".]

Lemma 1. If it cannot be proved that the Fermat four exist, then they do not exist.

* This search came to an end in 1995, when Andrew John Wiles [1953-] and Richard Lawrence Taylor [1962-] proved aspects of Taniyama-Weil Conjecture, entailing a solution of Fermat's Conjecture. See: <https://www.britannica.com/science/Fermats-last-theorem>

Remarks. Let A be a statement. There is no reason for thinking, that if it cannot be proved that A, then A is not true. However—and herein lies the essence of the lemma—that is the case, as soon as A is the statement that "the Fermat four exist".

Proof of lemma 1. We shall proceed from the opposite. Indeed, we shall assume that the Fermat four exist. Let us write out any of them—let it be the four natural numbers a, b, c, d. Let us verify that they really are the Fermat four, i.e. let us verify that the inequalities $a > 2$, $b > 0$, $c > 0$, $d > 0$, and the equality $b^n + c^n = d^n$, are satisfied. Presence of the four numbers a, b, c, d together with the indicated verification constitutes the existence proof for the Fermat four.

Lemma 2. If Fermat's theorem cannot be disproved, then Fermat's theorem is true.

Remark. There is no reason why this must be true of any theorem.

Proof of Lemma 2. Lemma 2 is simply a reformulation of Lemma 1. "To disprove Fermat's theorem" is "to prove that the Fermat four exist", and to assert that "Fermat's theorem is true" is to say that "the Fermat four do not exist".

The lemma 2, which we proved, has the structure "if P then Q". That is why, if P has a proof, then Q too has a proof (the proof of Q consists of joining the proof of the lemma with the proof of P). That is why, we have the following.

Corollary of lemma 2. If there exists a proof to the effect, that Fermat's theorem cannot be disproved, then there also exists a proof to the effect, that Fermat's theorem is true, i.e., simply put, a proof of Fermat's theorem.

In view of the importance of this corollary, let us Formulate it once more: *if there exists a proof to the effect that Fermat's theorem cannot he disproved, then Fermat's theorem can he proved.* Thus, if (b), then Fermat's theorem can be proved, and this is the promised negation of the statement (a_1).

The contradiction thus obtained concludes our arguments to the effect, that (a_1) and (b), and even more so (a), and (b), are incompatible.

Here arises the following natural question: but why these arguments cannot be repeated for the continuum hypothesis? Indeed, Fermat's hypothesis (theorem) states that the Fermat four do not exist, and the continuum hypothesis states that there exists no set having a cardinality intermediate to $\aleph_0$ and **c**. Now let us replace the Fermat four by a set of intermediate cardinality, and Fermat's theorem—by the continuum hypothesis and, let us once more adduce the arguments just adduced. We are bound to stumble somewhere, as the statements (a') and (b'), obtained from (a) and (b), by substituting the words "continuum hypothesis" for the words "Fermat's theorem", are both true. And where shall we stumble? Here's where—in the proof of lemma 1 (evidently, not in the initial formulation, but with the replacement of the words "Fermat four" by the words "set of intermediate cardinality"). The aforementioned proof of lemma 1 was based upon the following idea: that it is in fact possible to produce the four numbers a, b, c, d and to assure oneself that they are the Fermat four. But what does it mean to produce a set? Objections may be raised, that strictly speaking, we do not produce the numbers as quantitative categories, it is not possible to produce them, we can only write their names (for example, in the form of zero with the prime symbols or in the form of decimal notation).

However, the fact remains, that each natural number has a name, but such is not the case with the sets: there are more sets, than there are names (if we understand the latter as finite combinations of the symbols of some alphabet). But even if we limit ourselves to the sets having names, and produce in place of the sets—these names, there remains, all the same, a major difficulty: how to verify that the set produced has an intermediate cardinality? The verification to the effect that the four numbers are the Fermat four, is not complicated in principle (if we digress from the number of steps and the necessary space): we just have to put the numbers in the equation and compare the left hand side with the right hand side. But there exists no way of determining the cardinality of the produced set or of determining whether or not this cardinality satisfies the inequality $\aleph_0 < x < \mathbf{c}$.

The theme under consideration is most intimately connected with the famous incompleteness theorem of Gödel. This theorem states that *whatever be the proposed concept of a formal proof, there would be such a statement about the natural numbers, that neither it itself, nor its negation may be formally proved within the frame-work of the proposed concept*. We begin with the self-evidence of the fact that it is possible to define formal proof variously. These definitions differ from one another in respect of the collection of permissible axioms and the rules of deduction. It is possible to have such notions about formal proof, wherein there is no use at all, either of the axioms or of the rules of deduction. Briefly speaking, the approaches to the concept of formal proof may be very very different. But all these approaches have a fundamental generality expressed in the following principles:

1) every formal proof is a text—i.e., a finite chain of symbols, chosen from some alphabet;
2) in respect of every text, composed of the letters of an alphabet under consideration, it is possible to algorithmically identify, whether or not it is a formal proof, and if yes, then what, namely, does it state;
3) only true statements can have formal proofs.

On the strength of the third principle, the production of a formal proof of some statement guarantees its truth and, consequently, may be considered to be its proof. The converse, of course, is not being proposed: it is not being proposed that every true of even essentially provable statement has a formal proof, in terms of a pre-given concept of formal proof.

An analysis of Gödel's incompleteness theorem shows, that the statement therein discussed always has the form $\exists x\, \mathfrak{A}(x)$, where $\mathfrak{A}$ is some property of the natural number x. This property depends upon the concept of formal proof under consideration, but it is always algorithmically verifiable (just as it is possible to algorithmically verify the property of "being the Fermat four", in respect of four given numbers), [being algorithmically, verifiable means—there exists an algorithm, which verifies for any $\mathbf{c}$, whether or not $\mathfrak{A}\,(\mathbf{c})$ is true]. Thus Gödel's theorem states that neither $\exists x\ \mathfrak{A}\,(x)$, nor $\neg\exists(x\ \mathfrak{A}\,(x)$ has a formal proof.

Let us make our demands about the notion of formal proof even more strict. Let us demand, namely, that as soon as the statement $\exists\, x\ \mathfrak{A}$ turns out to be true for some algorithmically verifiable property $\mathfrak{A}$, then and there this statement $\exists\, x\, \mathfrak{A}$ possesses a formal proof. This demand is quite natural: it is realized upon formalization of the following steps indicated above: 1) the production of some c; 2) the verification that this C satisfies the property $\mathfrak{A}$; here it is essential, that c may in fact be produced and, that $\mathfrak{A}\,(c)$ may in fact be verified.

Our demand follows from two even more natural demands:

1) if the (algorithmically) verifiable property $\mathfrak{A}$ is valid for a number c, then $\mathfrak{A}\,(c)$ has a formal proof;
2) for any property $\mathfrak{A}$ whatsoever, if for some c the statement $\mathfrak{A}\,(c)$ has a formal proof, then the statement $\exists\, x\, \mathfrak{A}\,(x)$ also has a formal proof.

Now, with the help of arguments analogous to those used in connection with Fermat's theorem, we arrive at the following conclusion: if neither the statement $\exists\,(x)\,\mathfrak{A}\,(x)$, nor its negation $\neg\exists\, x\, \mathfrak{A}\,(x)$ has a formal proof, then from this information alone about the given situation it is possible to find out which of these two statements is true: namely, it is true that $\neg\exists x\, \mathfrak{A}\,(x)$.

Indeed, had it been true that $\exists x\, \mathfrak{A}\,(x)$, then this statement would have had a formal proof; perhaps it is not true that $\exists x\, \mathfrak{A}\,(x)$, and it is true that $\neg\exists\, x\, \mathfrak{A}\,(x)$ [the words "true" and "correct" are synonyms, but the word "provable" has another meaning (even other meanings)].

Let us appreciate the paradoxicality of the situation once more: *from the sole fact that neither A, nor not-A has a formal prof, it is possible to conclude, which of these two sentences is in fact true.*

6. What is a proof?

When we read a book written some fifty years ago, then the arguments found there, appear to us to be largely bereft of logical rigour.

Jules Henri Poincare, 1908.

(Nauka i metod, kn. II, gl. 2, § 4; [2, s. 356]).

In the previous reflection we came across the terms "proof" and "formal proof". It is sometimes thought that a formal proof is a proof that is formal. We would prefer to take a different look at these concepts.

A *formal proof* is a mathematical object, like, say, a matrix or a triangle. It is a finite chain of the symbols of some pre-fixated alphabet, i.e., as they say in mathematics, a *word* according to this alphabet. In the given instance, when we speak of a "symbol", we do not have in view the meaningful, contentful side, but only the external, graphic aspect of it is taken into consideration. To stress this circumstance, in mathematics, when the external, graphic aspect is had in view, then they speak not about a "symbol" [or "sign"], but about a "letter"'. Usually, the letters of the alphabets of various (Russian, Latin etc.) languages, numerals and the punctuation marks are considered to be letters. It would be reasonable, to consider the gaps among the words to be letters too (words in the ordinary, and not in the mathematical sense); we may devise some special symbol for it, for example #. This creates a possibility for viewing a text, i.e., a sequence of words, also as a word (in the exact mathematical sense indicated above). Thus, a formal proof is first of all a word in some alphabet—in the alphabet of formal proofs. It is clear, that this does not exhaust the concept of formal proof in the least: we simply wanted to stress that the concept of formal proof belongs to the class of words—just as the concept of triangle belongs to the class of geometrical figures.

What sort of words may be considered to be formal proofs? That is the theme of a special discourse; it is beyond the cycle of topics we wish to discuss here. We stress here that it is possible to give various definitions of the concept of formal proof, each of which would lead us to its own set of formal proofs. In the previous reflection we have enunciated some general postulates, to which any reasonable definition should be subordinated. It must be mentioned, however, that sometimes yet another step is taken in the side of generality and it is not demanded beforehand, that only true statements should have formal proofs, thereby the concept of formal proof is fully separated from the concept of truth. And afterwards this discarded requirement is introduced in the form of a supplementary property (which a formal proof, generally speaking, may not have): namely, if all statements having a formal proof are true, then the set of formal proofs is called *semantically non-contradictory*. A more precise general notion of formal proof is enunciated with the help of the concept of deduction; see, for example, [21].

We would like to stress once more, that not the contentfully understood statements themselves, but only their representations (i.e., again words) may (or may not) have formal proofs , in some precisely given logico-mathematical language.

The definition of the concept of formal proof—perhaps, it would be better to say: the definition of the sets of formal proofs—within the broad limits (conditioned by the general limiting properties of the sets of formal proofs, indicated above), happens to be arbitrary. Here, we have in view that "juridical" arbitrariness, which distinguishes mathematical definitions in general. For example, we have the "juridical" right to arbitrarily define a class of functions and to call it, "as we wish", say—continuous.

It is another matter, that any reasonable mathematical definition usually claims to correspond to some intuitive notions, to reflect them. Legitimacy of a definition still does not signify its reasonableness. Thus, the mathematical concept of a continuous curve reflects (with some sort of precision) our intuitive, contentful notions of the trajectory of a moving point. Analogously, the concept of formal proof reflects the intuitive notions of a contentful proof.

It may be said that the concept of formal proof is a mathematical model of the concept of proof—in the same sense, in which the concept of continuous curve is a mathematical model of the concept of trajectory.

It still remains to be explained: what a proof is. We have indicated at the very beginning of the present cycle of reflections, that it would be incorrect to assume that in mathematics everything is proved; however, there is no doubt about the fact that the concept of proof plays a central role in mathematics. "From the time of the Greeks, to say 'mathematics' is to say 'proof'"—thus begins Nicolas Bourbaki his "Éléments de mathématique" [6, p. 23]. At the same time we have noted that the concept of proof does not belong to mathematics (only its mathematical model—the formal proof, belongs to mathematics). It belongs to logic, to linguistics and, above all—it belongs to psychology.

Thus, one of the most important terms in mathematics, the term "proof", has no precise definition. An approximate definition of it is as follows: *a proof is a persuasive argument, which so persuades us that with the help of it we become capable of persuading others* [12].

Having grasped a proof, we become aggressive to a certain extent, ready to convince others with the help of the arguments which we have grasped. If we are not so ready, then it signifies that we are yet to grasp the presented argument as a proof, and even if we have given it the recognition of a proof, then we have done so simply to brush aside something.

We find that the concepts present in our definition of a proof are either logico-linguistic ("argument"), or psychological ("persuasive strength", "readiness") in nature. This fully meets the essence of the matter: the very notion of proof is inseparably connected with the linguistic means and with the social psychology of human society. And both of them change in the course of history. Linguistic formulations of proofs change. Our notions of persuation change.

The notion of persuation depends not only on the epoch, but also on the social surroundings. Unfortunately, I am unable to recollect now, where I read a passage on the following theme. The Cardinals of the time of Galileo, were quite intelligent, some of them saw with their own eyes the mountains on the moon through Galileo's telescope, and could follow the logic of Galileo's arguments. However, for them, their own views, based on an *a priori* dogma, were more convincing than any experiment and any logic . [In an article by S.P. Bozhich [13], we find an interesting analysis of how an *a priori*, predetermined notion about the ways of proving things prevents the recognition of certain facts.]

The notion about the persuasiveness of this or that argument depends on many factors. Revealing these factors happens to be an important task of logic and psychology. For example, the division of concepts (to be more precise, of terms) into sensible and senseless ones, happens to be one of these factors. The concepts of phlogiston and thermogen were considered to be meaningful in the 18th century, but they are now considered to be meaningless. Einstein discovered that the concept of simultaneity of two events is meaningless, as an objective concept independent of the observer (to be more precise, he discovered that simultaneity is not a two-place relation between two events, but a three-place relation involving the 1st event, the 2nd event, and the observer, as its terms). On the other hand, such an "evidently meaning-less concept" as the infinitesimal number, now fits into an exact meaning within the framework of a new branch of mathematics—the so-called non-standard analysis. With the changes in the notions about meaningfulness or meaninglessness of concepts, the notion about the very essence of scientific truth also changes. The notion of evidence too changes. Once everyone knew that "higher forces" unleash storms, now everybody knows that storms are caused by atmospheric electricity. In the case of inert gases, their property of not taking part in any chemical union was so evident, that this property was fixated in the very name "inert"; when, in 1962, these gases were for the first time found to take part in chemical unions, then, apparently, the chemists were not ashamed, rather—they happily stated that "for explaining the structure of these unions, we did not need any, in principle, new notion about the nature of chemical bonds" (The Great Soviet Encyclopedia, 3rd ed. the article on the "Inert Gases").

It is common ground that human knowledge changes with the march of history. Here one would like to stress that not only the facts themselves go into the composition of knowledge, but also the initial positions and presumptions, on the basis of which this or that fact becomes a component of a system of knowledge : these are the notions about meaningfulness and meaninglessness, about obviousness and non-obviousness, of the possible and the impossible, of the part and the whole, persuasiveness and the lack of it, the proved and the unproved and, the authentic and the inauthentic. It is possible, that all these notions change more slowly than the simple notions about facts, but in essence, they are historically as relative as our notions about facts.

Mathematics is sometimes perceived as a stationary rock towering above the waves of changing notions about the other disciplines. Of course, there are grounds for such a view of mathematics. At the same time, the notion of some absoluteness of mathematics is evidently exaggerated. If mathematics is absolute, then it is so only at the level of everyday experience—just as Newtonian physics is absolute in its application to the phenomena of "medium size" (and yet another—Einsteinian—physics operates at the level of the small and the very big) [see the already mentioned article of P.K. Rashevsky [16]: on "spreading in a large" representation notions of the natural number].

In particular, the socio-historical conditionalities of the notions of proofs are on the whole extended upon mathematical proofs.

To illustrate what has been said above, I shall now briefly narrate my understanding of the concept of proof in ancient Egypt, ancient Greece and in India.

We do not have much authentic information as to how mathematical proofs were enunciated and understood in the ancient period. The texts that came down to us are in many cases fragmentary: moreover the terms contained therein often have debatable interpretations; for example, on the interpretation of ancient Egyptian mathematical texts, see the remarks of the translator in: [4, p. 139]. There is a lot that is conjectural. Every one makes conjectures in the direction s/he wishes, and the present author is no exception. Taking these stipulations into account, the following outline may be proposed.

The proposed outline is based on the conviction that the notion of proof is a product of the history of societies. We are aware of the simplification involved in our historical excursus, as we describe ancient Egypt as a centralized state—since there have been periods of splintering there, or ancient Greece—as a democracy, since there, too, there have been cases of tyranical or oligarchic rule. But then, any outline involves some simplification.

Ancient Egypt. A centralized theocratic state, with an extraordinarily strong discipline. Continuous construction of pyramids—requiring colossal human and material resources and uniting the strength of the entire land—served as an effective instrument for maintaining centralization, discipline and order. Authority of the Pharaoh and of the priests was incontestable. Authority of the written word was also unquestionable. If a priest, scribe or teacher said or wrote something, then that means—such is the case. If something is written on papyrus, then that is the case. Persuasiveness was based on the authority of the source.

Ancient Egyptian mathematical texts contain ready-made recipes without any substantiation. When we speak of absence of substantiation, here we have in view the modern understanding of the word "substantiation". From the point of view of a person of that time a recipe on a papyrus was fully substantiated, as it came from an authoritative source and was drawn up in the authoritative form of a record on papyrus. The fact of being recorded on a papyrus, was in itself the proof. In reality, this fact was enough for convincing others with the help of it. A number of recipes for computing the areas of triangles and quadrangles have been non-univocally interpreted in our time; the disputes about how to understand the terms contained in these recipes, still continue [4, ch. IV, § 2, a]. Depending upon these interpretations, these formulae may be taken to be either exact, or approximate, or totally incorrect. When we speak of incorrect formulae, here, we have in view the representation of the area of a triangle through half of the product of the base and a side of it. This is what academician L.S. Pontryagin has to say on this score: "The first mathematical manuscript known to us—is the manuscript of Ahmes, composed some 2000 years before our era. It contains some algebraic and geometrical rules—for example, for computing the area of a triangle ... However, the Ahmes Papyrus contains a mistake. According to him the area of an isosceles triangle is equal to the product of its base and half of a side—but to-day every school-student knows that it is not true" [25]. However, many a researcher thinks that the corresponding ancient Egyptian term should not be translated as a side, it should be taken to mean height (and then the formula contained in the papyrus turns out to be true. However, even if this term did in reality signify, not the height, but a side, the corresponding (according to our modern point of view incorrect) formula should be considered as proved according to the ancient Egyptian understanding of the word "proved": as this formula is convincingly substantiated by the fact that it (of course, not as a formula, but as a recipe expressed in words) is contained in an authoritative document.

The situation was somewhat different in *ancient Greece*. (In comparison to Egypt) here we have comparatively small state formations together with popular assemblies. The orators, who spoke in these gatherings did not carry any *a priori* authority. They had to convince the listeners by arguments. Formulating correct arguments became an everyday and actual requirement. Hence the birth of logic in the hands of Socrates, and its final shaping as a discipline by Aristotle. Hence also, the beginning of the deductive method in mathematics, approaching the modern notion of proof. Arguments became the basis for mathematical conviction. The concept of the foundations of correct arguments, of axioms and postulates, arose. That which could be obtained through "legitimate arguments" from the initial statements, considered to be valid, was considered to be convincing (and consequently, proved).

Finally, *India.* We intend to refer to some geometrical figures, taken from mediaeval Indian texts, but that does not mean that these figures did not appear in ancient India. In general, the task of dating of Indian mathematical notions gives rise to considerable difficulties, as some texts may be expositions of some other earlier texts. On the other hand, it is not so essential either: mediaeval Egypt and Greece had nothing in common with ancient Egypt and Greece, but mediaeval India remained the custodian of the intellectual heritage of ancient India. An essential trait of this tradition was (and is) the conferring of the status of highest authenticity to the inner light. Immediate inner illumination was considered to be the basic source of knowledge and it had indisputable persuasive power. That which was thus known was considered to be proved. In order to convince others of it, those others must be brought to such a state, that they themselves experience the inner illumination. That is why, a geometrical proof had two parts: a diagram, and below it the inscription—"See !".

We find the examples of such diagrams with the inscriptions "See !" in some texts dating back to the 12th-16th centuries [9, p. 76 and 154]. We reproduce below one of these diagrams; it has also been reproduced in: [15, p. 75]. We are of the opinion that it deserves to be included in any modern, secondary school level, textbook of geometry: it shows, more graphically than the modern proofs, that the area of a circle is equal to the area of a rectangle , the two sides of which are respectively equal to half of the circumference and half of the diameter of the given circle. See figure 5 below.

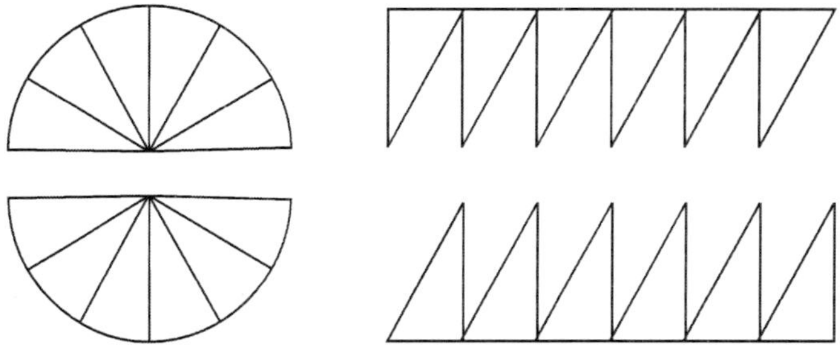

Fig 5

The present author is aware of the fact that his views on the Indian proofs are different from those of an authority in the field of history of mathematics, like A.P. Yushkevich, who wrote: "The laconism of inferences in the Indian works on mathematics or the presence in them of diagrams together with only the inscription "See !", should not be viewed as the manifestation of some, special approach to the problem of proof or of some special movement of thought" [9, p. 155]. We are of the opinion that they should be so viewed. Or else, why do we not come across this kind of "See !" anywhere else? Why only in India?

S.S. Demidov has put forward valuable considerations on the evolution of the concept of mathematical proof in [15], where, in particular he states that, "in the final count, the proof (giving power) of mathematical arguments is their persuasive power. What appeared to us to be convincing yesterday, does not appear to be so to-day".

The definition of proofs as convincing / persuasive texts makes the concept of proof very subjective (for some, a text is convincing, for others—not). We do not consider it to be a deficiency of the definition. Such is the state of affairs. Perhaps, the use of the word "makes" above has been unfortunate. Our definition *does not make* the concept of proof subjective, it *only reflects* the subjective character of this concept. Even more interesting is the problem (we are very far from solving it), as to why, nevertheless, the concept of proof has an universal-cultural character in the sense, that within the limits of one and the same culture, though there occur disputes about whether or not this or that statement is true—such disputes are comparatively rare.

While speaking of such disputes, we do not have in view the disagreements among the representatives of various logical trends in mathematics, for example, those among the representatives of the ordinary, classical mathematics and the representatives of the intuitionistic (constructive) mathematics. The latter do not recognise many statements of ordinary mathematics as proved (and, on the contrary, consider them to be untrue). It may be said that the intuitionists and constructivists belong to different mathematical cultures and even the most customary words (like, say, "exists") has a different meaning for them (evidently, the intuitionists and the constructivists think that the representatives of traditional mathematics put different meanings into words, and it is they—the intuitionists—who use these words in their only correct meaning). That is why the intuitionists consider many proofs of traditional mathematics to be invalid.

Here we are talking of something else—not about changes in the semantics of terms, leading to the changes in the truth values of statements, but about the fact that a proof may turn out to be not understood and that is why not convincing (and once not convincing—it is not a proof at all). Modern mathematics has a complex structure, which has almost stopped to be visible. The proofs of some of the theorems turn out to be so cumbersome, that in order to be able to verify them one must have an extraordinarily big desire, patience and time, to say nothing of the fact that one must have special knowledge—for a number of theorems, not only the invention of their proofs, but even the verification of these proofs appear to be accessible, only to a narrow circle of refined specialists.

Sometimes the volume of the proof of this or that theorem becomes an object of interest. Here, we often find that some theorems established earlier—which are no more required to be proved—are permitted to be used as ready-made formulations in a proof. Will such an argument be a proof—i.e., a convincing text—for someone who does not know the proofs of these theorems "established earlier"? We do not intend to give an univocal answer to this question. We would like to mention further, that the very word "earlier" introduces an additional subjective"relativistic" moment (two almost simultaneously proved theorems may be chronologically differently ordered by different observers). If any reference to any theorem whatsoever proved earlier, is forbidden in a proof and if one is required to go back directly to the definitions and primary, undefined concepts (which we have discussed in our first reflection), then such a complete proof, may, in a number of instances, stretch into thousands of pages of mathematical text (and may even be more difficult to perceive, than a proof based upon clearly formulated—though not known to the reader—facts).

The study of difficult mathematical proofs may be compared with a mountain-climber's ascent to the peak. The sea-level corresponds to the initial concepts. Ascent from the sea-level may take months, and its mathematical analog (understanding a proof) may take years. In both the cases, there are many intermediate stops. First, you go to the common high-altitude base camp, here the climbers going to the various neighbouring peaks gather. This stage corresponds to the stage of serious mathematical preparation, sufficient for acquiring an understanding of the more special themes. Then begins the assault to the chosen peak; here again, we have intermediate camps and stops. For mathematics, the corresponding theories and theorems play the role of these camps and stops. Just as a mountaineer may make a limited number of ascents in his life time, so may a mathematician—get to know only a limited number of proofs.

The following common trait of the mountaineer and the mathematician is an important one—there is some kind of conventionality involved in the choice of the point of departure. The ascent proper does not begin at the sea-level, but from a point, where the professional mountaineers may be able to gather without difficulty, but it may be a matter of great difficulty for an ordinary person, if s/he wants to arrive there. The proof proper begins from an analogous point: this point is situated at some general-cultural (we have in view the mathematical cultural) level. However, at the present stage of mathematics the generality of the prefix "general-" is being lowered continuously, and now many proofs begin form a point, accessible only to the narrow specialists. A second common trait consists of a break up into stages, the presence of sufficient number of intermediate stops (camps).

Where from does one get the conviction in mathematics, that the proved theorems—the proof of which one never gets to know—are indeed proved, i.e., have proofs? Evidently such conviction is based on trust alone. Seen from outside, such a situation should not appear to be very strange. Indeed how many of the readers of these lines have seen the Easter Island? For those who have not, the conviction that this island exists, is also based, in the final analysis, on trust. But if a modern proof is based upon trust in authority, then how is it, in principle, different from the ancient Egyptian proof?

It is not a simple question. Perhaps, the answer to it lies in the fact that proofs are gradually moving over from the ranks of the phenomena of individual experience to those of the phenomena of collective experience. Pushing the collective to the fore is in general characteristic of the history of civilization. It is well known (and widely discussed) that with the development of human society, there arises division and cooperation of labour and, this gets strengthened steadily. Only in the deep antiquity could man himself produce all that he needed: now everybody is required to use the results of the labour of others. It is known (though less discussed) that division and cooperation in scientific knowledge takes place simultaneously. It is difficult to say when—perhaps in the middle ages—one could have found individual scholars, capable of grasping the totality of the knowledge of his time. Now everybody must, this way or that, use the knowledge of others. The situation with proofs is an analogous one: the activities in the sphere of production and consumption of proofs have become as much an object of division and cooperation of labour, as are the activities in the sphere of production and consumption of knowledge. The very concept of conviction has begun to lose its individualized nuance and, is more and more assuming the character of "collective conviction". Evidently, we must gradually learn to speak of the conviction, of not a separate individual but, of some scientific collective. Here , collective conviction does not in any way signify that it is equal to the "immediate conviction" of each of the individual members of the collective. A collective does not function as a simple totality of its members, but as a single whole. The idea of collective conviction indicates the fact, that for every component part of a proof we have a member of the collective "answerable for it", who is immediately convinced, namely, about that part (and other members of the collective rely on this member on that given question).

The age of informatics is introducing its own correction, also to the notion of proof. For instance, there arise situations when a proof involves sorting out of such a large number of variants, that it becomes difficult for a human being to do the sorting, but a computer can do it. Let us assume that a computer sorted out all the required variants, and the sorting led to the necessary results. Can we then say, that we have obtained a proof? And what if the machine "malfunctioned"? (But a person also makes mistakes!) That apart, a guarantee is essential to the effect that the programme was right; correctness of programmes can be ascertained only with special proofs, and the theory of such proofs constitutes a special division of the theory of programming.

In reality, computers were used to solve the four colour problem. [For a formulation of this problem, see: the article "The four colour problem", in the 3rd edition of the Great Soviet Encyclopedia.] In terms of simplicity of formulation, this problem, consisting of a proof of the four colour hypothesis, is hardly inferior to Fermat's problem (consisting of a proof of Fermat's hypothesis), but in terms of the naturalness of the statement (and the applied significance) it is superior to Fermat's problem. The solution of this problem was announced by Appel and Haken in 1976 [17] and set forth in 1977 [18 and 19]. This solution is based upon a reduction of the solution into a large number of particular instances, the study of which was entrusted to computers. The computers verified all of them, and thereby it was proved that every map is four colourable, as per requirement.

Appel and Haken themselves said about their proof [20]: "The proof involved an unprecedented use of the computers. The calculations used in the proof made it longer than what is traditionally considered to be permissible. In fact, the validity of the proposed proof can never be verified without the help of computers. What is more, some of the decisive ideas of this proof materialized through computer assisted experiments. It is possible, of course, that one fine morning there would appear a short proof of the four colour theorem ... At the same time, it is also conceivable that such a short proof is not at all possible. In the latter case, new and interesting types of theorems emerge, for whom traditional types of proofs do not exist".

Of late, however, the validity of the proof provided by Appel and Haken came to be doubted. The doubt is not about the computer-use part of it, but about the pre-computer, theoretical part—wherein it is sought to be established that the entire problem is really reducible to a consideration of the particular instances.

Commentary. Let us describe the situation involving the proof of Appel and Haken in somewhat greater detail, The basic idea of its authors is connected with the following notions.

First of all, the authors go over from the colouring of the regions in a map to the colouring of the apexes in a planar graph, such that it is a triangulation. Further, they call any sub-graph, forming a cycle, and the interior of that cycle—a configuration. If it can be proved by some standard method, that a configuration cannot be immersed in a minimal counter-example of the four colour hypothesis, then it is called—*reducible*. If each planar triangulation contains one of the configurations of a set as a sub-graph, then that set of configurations is called unavoidable. From these definitions it easily follows that for obtaining a (positive) solution of the four colour problem it is enough to produce an unavoidable set of reducible configurations. The authors of the proof produced 1834 explicit, reducible configurations, forming the unavoidable set [19. pp. 505-567]. In each of these configurations, the length of a cycle was 14 or less. Computers were used both for finding the unavoidable set and for proving the reducibility of its terms.

If in the first case (construction of the set) the computers were drawn into a helping role, since the very proof of unavoidability of the set obtained (now it is not important, how that was done) is not based on computer-assisted-calculations, then in the second case (of verification of reducibility) the use of computers happen to be an essential component of the proof. and each configuration needs some 10 minutes of computer time for such verification. While evaluating the proof of Appel and Haken, some reviewers indicated [23] that the authors of the proof needed four years and 1200 computer hours for its construction and, that the text of the proof takes 139 pages, including 99 pages of drawings, the average size of more than 30 of these drawings being one page. The reviewers commented that the "essentially search type character of the proof makes its verification difficult (according to Appel the verification of all the details requires 300 computer hours)". Evidently, these 300 hours are required for the verification of reducibility. However, as we have already mentioned, the non-computer part of the proof—involving the verification of the unavoidability of the set of configurations produced—gives rise to doubt. The fact remains that the text of the proof [18 and 19] does not present this verification directly and exhaustively. We have been informed through a foot-note on p. 460 that the details of the proof of unavoidability of the presented set (to be more precise, details of the proof of the so-called spacing out theorem, which provides the basis for this unavoidability), are contained in the microfiches, supplied as a special supplement to the journal. However, the present author could not go through that supplement.

It seems that with the development of mathematics (and with the appearance of ever more complex and long proofs) the proofs are losing an important trait—that of being convincing. One fails to understand, then what remains of the proof: conviction / persuasiveness enters into the very definition of proofs! That apart, with the growth in the complexity of proofs, their element of subjectivity grows too. Of course, a formal proof is objective. But, firstly, not the judgements themselves, rather their expressions, their representations in formal languages, that have formal proofs. Secondly, though the verification of the statement, that a given text is a formal proof, is accomplished algorithmically, it may give rise to considerable difficulties, in the case of a voluminous text.

Large proofs begin to live by some macroscopic rules. Just as the concept of natural number gets diffused in the case of the "large" numbers (once more we refer the reader to P.K. Rashevsky's article) [16], so does our notion of a proof; it gets diffused, when the volume of a proof becomes inordinately large.

It so happens, that though all proofs should, by definition, be convincing, some proofs are more convincing than the others, i.e., as though, some happen to be, to a greater extent proofs, than are the others. There emerges something like a gradation of proofs according to the degrees of demonstrability—of course, such an idea fundamentally contradicts our primary notions about the identical indisputability of all proofs. But mathematical truths do permit a gradation of that kind. Each of the three following statements—"$2 \cdot 2 = 4$", "$17^{14} > 31^{11}$" and "$300\,! > 100^{300}$"—are true. However, we say: "True as $2 \cdot 2 = 4$", and do not say: "True as $17^{14} > 31^{11}$" or "True as $300\,! > 100^{300}$".

7. Can mathematics be made understandable?

Why so many people do not understand mathematics? The great Poincare was disturbed by this problem and, he wrote: "How to explain, why many an intellect refuses to understand mathematics? Is it not paradoxical? Indeed.... here we have a problem, that does not lend itself to an easy solution; all those who wish to devote themselves to teaching must take up this problem" [2, p. 353].

Most probably both the sides involved "are to blame". Non-mathematicians are to blame: a bad education has trained them into a non-understanding of and even into taking a hostile altitude towards mathematics (as Poincare has noted "often the intellect of those people who are in need of guidelines, is very lazy to seek them out") [2, p. 354]. Mathematicians are to be blamed: they do not wish to waste their strength, explaining their mathematics to the uninitiated (and how many people are astonished to find, that there still remains something to be discovered in mathematics!). Of course, in mathematics there would always remain numerous details inaccessible to the non-professionals (and even to the professionals, of a different field of mathematics). But such is the case everywhere—for example, in chess, even the other grand masters do not understand many a move, when Karpov and Kasparov battle against one another. At the same time, a very large part of mathematics, larger than what is usually thought to be the case, may be explained to a wider circle of well-meaning listeners and readers—of course, not in detail, but at the level of the herart of the matter. Clearly, this would require that the mathematicians engage themselves single-mindedly in this new direction of activity. Perhaps, thereby they would be discharging their moral duty to the humankind.

"But in order to help those who do not understand, first of all, we must know what restrains them" [2, p. 345]. It appears, that the complex logical structure of mathematical definitions and statements, in which the logical connectives and the existential and universal quantifiers take turns, happens to be the hindrance in many cases. Every teacher of mathematical analysis knows the difficulty that arises in the course of parallel assimilation of the concept of limiting point of a sequence—the definition of which has the structure

$$\forall\ \varepsilon\ \forall\ k\ \exists\ n\ (A \wedge B),$$

and the concept of limit of a sequence—the definition of which has the structure

$$\forall\ \varepsilon\ \exists\ n\ \forall\ k\ (A \Rightarrow B).$$

However, are these psychological difficulties encountered by the learners, while assimilating these concepts—difficulties pertaining to the heart of the matter or, are these difficulties of linguistic expression? I do not have any final answer to this question. It is connected with an even deeper question: is it possible to separate mathematics from its linguistic formulation? In other words, does mathematics abide exclusively in the mathematical texts or does mathematics have some other essence, different from the texts—and the texts serve only as this or that (and perhaps not always felicitous) mode of expression for that essence. It is clear that this question, which we have called a "deeper" question, is applicable not only to mathematics, but also to any other discipline. According to a formulation of Engels, mathematics is different from the other disciplines in so far as it is "an abstract science, dealing with intellectual constructs" [1, p. 529]. [These intellectual constructs can hardly be understood by the human intellect, if they are not based on ordinary human logic, and consequently—on reality, from the operations with which, this logic has come into being.]

Like all rational concepts, the mathematical concepts too exist in the form of notions, not necessarily connected with texts. The linguistic texts defining these concepts should be recognized as important, but not as the only, means for their assimilation.

It appears, that now we have at our disposal more adequate means of introducing the concepts of limit and limiting point of a sequence, to those learners (who do not have special "mathematical capabilities"—that is, according to modern understanding, to those who do not have a high capability of assimilating, namely, linguistic formulations). Let us imagine a screen, on which we may draw the trajectory of the movement of a point, unboundedly approaching some other stationary point, which is the limit. This has to be repeated a number of limes, with changes in the position of the limit (so that the false impression is not created, that every sequence has one and the same limit), as well as in the mode of approach of the moving point to the limit (so that, in particular, the false impression is not created, that the distance between the moving point and its limit changes monotonically). It is possible to present an analogous graphic illustration of the concept of limiting point: when, though the trajectory unboundedly approaches that point at times—at others, it moves away from it by a definite distance. It appears very likely, that any viewer of such pictures would form a correct notion both of the limit and the limiting point.

One is led to believe that with the introduction of computers, teaching will proceed along the path of visualization of concepts, traditionally considered to be entirely abstract.

Had the theme under consideration been one of pedagogical significance alone, then we would not have dwelt upon it so elaborately in an essay of philosophical character. However, this theme exceeds the bounds of pedagogy and, closes up to the question of ontological nature of mathematical concepts. Like all other rational theoretical question, this question too has an applied significance—in the given case, in the order of reverse connection, it is pedagogical. Indeed, if a mathematical concept has an essence, different from its embodiment in a linguistic definition or formula, then one can hope for a better understanding of that essence, by demonstrating its various manifestations (and not only its formulation).

In order to adduce a proof, we shall consider a fresh example. On pp. 71-72 of a recently published textbook [24], there is a formula that defines a mathematical concept—the so-called Clarke's cone. Having formulated its definition the authors wrote: "However, at the first glance, it is neither possible to understand the properties of Clarke's cone, nor the meaning of its formal definition itself". And further on, they have at first put forward some heuristic arguments explaining Clarke's cone, and then translated these arguments in the language of non-standard analysis. Here one gets the idea that as though the concept of Clarke's cone exists all by itself; its definition in the form of a formula is only one of the means (and not the most felicitous means) of comprehending this concept, but descriptions like the "results of examination of the set through a microscope" [25, p. 86], are useful for a better understanding of it.

Independently of the fact, whether or not such is indeed the case, we may put forward the following fruitful working hypothesis: a truly "profound mathematical concept or mathematical statement must in essence be simple. And then there is a hope, that it will be understandable (or better still, understood): it is easy to get used to that which is simple, and we do not know any interpretation of "to understand", other than "to get used to".

Literature

1. *Marks K., Engels F.,* Coch. 2-e izd. T. 20.
2. *Poincare H.,* O nauke. M., 1983.
3. *Hilbert D.,* The Foundations of Geometry. Chicago 1902 [Russ. tr. 1948].
4. *Neugebauer O.,* Vorlesungen über Geschichte der antiken mathematische Wissenschaften. Erster Band: Vorgriechische Mathematik. Berlin. 1934 [Russ. Tr. 1937].

5. Tolkovy Slovar Russkovo Yazyka. M. 1938. T. 2.
6. *Bourbaki N.,* Theory of Sets. Mass., 1968. [Russ. Tr. 1965].
7. *Church A.,* Introduction to Matematical Logic. Princeton, 1956 [Russ. Tr. 1960].
8. *Hornby A.S.. Parnwell E.C.,* An English-reader's Dictionary. Oxford. 1959.
9. *Yushkevich A.P.,* Istoriya mathematiki v sredine veka. M., 1961.
10. *Pototsky M. V.,* O pedagogicheskikh osnovakh obucheniya matematike. M., 1963.
11. *Gorsky D.,* Opredeline // Filos. Entsiklopedia. M., 1967. T. 4. c. 150-152.
12. *Uspensky V.A.* Predislovie // Mateinatika v sovremennom mire. M., 1967.
13. *Bozhich S.P.* O sposobakh istinnostnoi otsenki estestvennonauchnovo vyskazyvaniya // Logika i empiricheskoe poznanie. M., 1982.
14. Izomorfizm // Bolsh. Sov. Enls. 3-e izd. M., 1972. T. 10.
15. *Demidov S.S.,* K istorii aksiomaticheskovo metoda // Istoriya i metodologiya estestvennykh nauk: Matematika. Mekhanika. M., 1973. Vyp. 14.
16. *Rashevsky P.K.,* O dagmate naturalnovo ryda // Uspekh. mat. nauk. 1973. T. 28. Vyp. 4(172).
17. *Appel K., Haken W.,* Every Planar Map is Four Colorable// Bull. Amer. Math. Soc. 1976. Vol. 82. N5.
18. *Appel K., Haken W.,* Every Planar Maps is Four Colorable . Pt. I: Discharging // III. J. Math. 1977. Vol. 21, N3.
19. *Appel K., Haken W., Koch J.,* Every Planar Map is Four Colorable. Pt. II: Reducibility // ibid.
20. *Appel K., Haken W.,* The Solution of the Four-Color-Map Problem // Scientific American. 1977. Vol. 237, N4.
21. *Uspensky V.A.,* Teorema Gödelya o ncpolnote. M., 1982.
22. *Plisko V.E.,* Teorema // Mat. Ents. M., 1985. T. 5.
23. *Kozyrev V.P., Yushmanov S.V.,* Teoriya grafov: (Algerian., algebraich., i metr. probl.) // Teoriya veroyainosti: Mat. Siatistika. Teoret. kibernetika. M., 1985. T. 23.
24. *Kusraev A.G., Kutateladze S.S.,* Subdifferentsialy i ikh primeneniya: Ucheb. Posobie. Novosibirsk, 1985.
25. Uroki otkryvaet beseda s matematikom L. Pontryaginym: Inierviyu akad. L.S. Pontryagina "Uchit. gaz. "// Uchit. gaz. 1985. 23 maiya.

Source: Zakonomernosti razvitya sovremennoi matematiki. "Nauka". M., 1987. s 106-155.

About the author: Uspensky, Vladimir Andreivich (1930-2018), Soviet and Russian mathematician, logician, linguist and publicist. Fields of specialization: theory of algorithms and mathematical linguistics. See: <http://lpcs.math.msu.su/~uspensky/index.htm>

Other works:

1. O ponyatii algoritmicheskoi svodimosti (1953);
2. Teorema Gödelya i teoria algoritmov (1953);
3. Ob algoritmicheskoi svodimosti (1956);
4. Ponyatie programmy i vychislimyie operatory (1956);
5. Uporyadochennye i chistichno uporyadochennye mnozhestva (1956);
6. Neskoilka zamechanii o perechistimykh mnozhestvakh (1957);
7. K oprcdeleniyu chasti rechi v teoretiko-mnozhestvennoi sisteme yazyka (1957);
8. K opredeleniyu padezha po A.N Kolmogorovu (1957).

Emergence and development of the concept of constructivisability in mathematics

Nikolai Nikolayevich Nepeyvoda

Constructivisability of a mathematical theory signifies the possibility of isolating the constructions of objects from their existence proofs.

Pre-Greek empirical mathematics was constructive by its very nature. It was preoccupied, namely, with the means of construction of objects, and gave empirical recipes in certain situations. Mathematical reasoning was reduced to one or, in the extreme cases, to several applications of such recipes, and the only descriptive element in it involved judging whether or not the problem at hand or, part thereof, belongs to a certain class. This descriptive element was most often reduced to an appeal to immediate obviousness.

Construction of the object being sought was the only method of proof in the Indian and Chinese mathematics, and this construction was able to take the place of arguments (we recall the famous Indian diagrams with the word "see"). Arguments could only help in construction, they did not have any independent significance [1].

The concept of proof came to occupy a proper place in Greek mathematics. Classical logic was used with all its might. It has been established in the 20th century, that this logic was suitable, in the first place, for describing the static universe of ideal concepts, and not for carrying out intellectual constructions. Though Aristotle did highlight the special logical status of the rule of contraries: "... one of these [direct proof] proceeds from the previous [knowledge], and the other [from the opposite] from the subsequent" [2, p. 307]—this remark, which astonishingly exactly reflects the semantics of the rule of contraries in Kripke's models, did not exert any influence upon strict mathematical arguments.

Nevertheless the use of classical logic—an instrument, oriented towards descriptive and not constructive applications—did not lead Greek mathematics to non-constructive methods and theorems. As before, existence proofs included (as a rule, geometrical) constructions. Arguments from the contraries were used only to substantiate the constructions already carried out, in the main, for proving the equality or inequality of certain magnitudes.

The deeper reasons behind this phenomenon were revealed only in the last few decades. It is intimately connected with the hold-up of the Greeks in front of the concept of real number, from the point of view of traditional mathematical paradigm—with their strange antipathy to the explicit use of numbers in general, in strict mathematical arguments. Vexations regarding the specificities of Hellenic mathematics—which are indeed not quite understandable from the classical point of view—have been expressed more than once. In particular, the question arises: why the real numbers were used in a masked manner, as proportions, and why acquaintance with incommensurability did not come in its way, and yet in geometry, the natural numbers were avoided in all possible ways, though, one would think, that these are sufficiently intuitively reliable objects? Why, the Hellenic arithmetic remained something like a handicraft or an art, never entering into the sphere of operation of "pure mathematics", save in the case of a few theorems like the one about the infinite set of prime numbers? What prevented the Greeks from formulating and utilizing such a powerful principle of conducting arguments, as the mathematical intuition?

It has been proved, namely, that the classical geometry and the elementary theory of real numbers are complete and solvable; see, for example, [3], Thus, for every concrete, closed statement in the language of these theories, it is provable in them, that it is either A, or $\rceil A$. Consequently, classical logic cannot lead us to the non-constructivisability of the theorems proved either in geometry, or in geometry supplemented with algebraic operations on the real numbers, but without the explicit mention of the integers as a set. In any classical proof of these theorems one may mark out the construction and its substantiation, which may be carried out, in particular, also by the method of "indirect proof".

This fact once more confirms the depth of the intuition of the Greeks, which was based upon purely aesthetic and methodological considerations, but which permitted them to stop, namely there, where the rupture between argument and construction, between descriptive and constructive knowledge, became important; Such an exact halt was conducive to the fact that, the distinction between what was constructive and what was descriptive was not realized and, was correspondingly erased out of the world outlook of mathematicians. Perhaps it gave an indirect push to the courageous introduction of numbers and their functions in the mathematics of the modern times: mathematicians were still unaware of the danger of a rupture between the proof and the construction, it was erroneously accepted that [the verbs] "to prove" and "to construct" were always mutually concordant. Consequently, as before, mathematicians assumed—now, without any foundation, simply due to inertia—that a strictly proved statement provided the means for the construction of those objects, whose existence has been affirmed. When the construction was explicitly indicated in a proof, then that was, of course, rated somewhat higher, but the pride of place was reserved for the other factors, in the first place—for the not explicitly formulated, and that is why constantly implicitly changed, aesthetic ones.

Prior to the formulation of the axiom of choice by G. Cantor and E. Zermelo, mathematicians did not realize that even after the explicit introduction of the totality of natural numbers together with the principle of mathematical induction, there would appear non-constructive theorems of existence, which would not provide the construction sought—even in principle. The axiom of choice is demonstratively ineffective. It states that, it is possible to construct a function, by choosing its elements from among each of the members of the family of non-empty sets, without saying anything about the method of carrying out this choice. The shock generated by the axiom of choice and by the paradoxes of the theory of sets—which appeared practically at the same time, forced the realization that a very large part of mathematics of the period ending in the 19th century was indeed non-constructive. The axiom of choice was magnificently inscribed upon the entirety of the hitherto formed paradigm of classical mathematics.

It should be mentioned here, that even in the 19th century attempts were made to construct some sections of mathematics upon a more constructive foundation—in particular by R. Grassmann [4] and E. Schröder [5]—but these attempts remained on the sidelines, away from the main road, and were forgotten.

Thus, the "crisis of the foundations of mathematics" sharply posed the question about the nature of mathematical constructions and about the interrelationship of mathematical objects and reality. And this gave rise to the necessity of a more exact and, in any case, a more explicit characterization of the class of constructive methods and, if possible, of eliminating the explicitly non-constructive ones. While examining the problem of constructivisability, it is possible to mark out three major trends: pseudo-classical, non-classical and significative.

The aim of the pseudo-classical trend is to single out the constructive sub-languages of the classical theories, wherein the classical logic and the customary concept of truth are left untouched. In modern mathematics this approach begins with H. Poincare—who tried to isolate those results of classical mathematics, which were obtained without the help of the axiom of choice, as he considered them to be more reliable and, with D. Hilbert—who had a more radical programme. For a detailed analysis of Hilbert's programme from a point of view that corresponds to the present state of the investigations in the foundations of mathematics, see: *Ershov. Yu. L. and Samokhvalov K.F.*, O novom podkhode k methodologii matematiki // Zakonmernosti razvitiya sovremennoi matematiki ("Nauka", M., 1987), pp. 85-106. Here we shall limit ourselves only to the remark that D. Hilbert unequivocally declared that the majority of mathematical statements do not have any real meaning and, that mathematics is required to give correct results only in respect of a set of comparatively simple real statements.

In the non-classical trend the concept of effective method is considered to be of paramount importance—mathematics is viewed as the science of effective (intellectual) constructions and logic adapts itself to the methods of such constructions, and gets so modified, as to wittingly guarantee the constructivisability of the constructions.

L.E.J. Brouwer was the first to point out that while aiming at attaining constructivisability one must not blindly follow that logic, which is tied to the tradition [6]. The roots of the non-constructive structures, are often not so much mathematical, as logical. For-example, in any recursively axiomatizable non-contradictory classical theory containing arithmetic, it is possible—basing oneself upon a theorem of Gödel—to construct a statement of the form $\exists x \in N A(x)$, such that it is not possible to construct even one such number n, that $A(n)$ is provable, but, nevertheless, $\exists x \in N A(x)$ is provable. Indeed, for this it is enough to take the statement A—which is unsolvable in the given theory, and to construct the formula

$$\exists x ((x = 0 \,\&\, A) \vee (x = 1 \,\&\, \neg \mathrm{A})). \tag{1}$$

Brouwer showed, in particular that, the following two logical principles are most open to criticism from the point of view of constructivisability: the law of excluded middle $A \vee \neg A$ and the method of indirect proof $\neg\neg A = A$. Indeed in the constructive substantiation of the law of excluded middle it is demanded that a general method be constructed in respect of every problem, for establishing whether or not a given statement is true, and in the majority of cases such a method does not exist. Thus, the law of excluded middle may be called "the principle of omniscience", and it may be applied only in that situation, where both the language and the interpretation are deliberately so selected as to exclude the possibility of emergence of unsolvable problems.

Analogously, the principle $\neg\neg A \Rightarrow A$ signifies, that there exists a method for transition from the formulation of a solvable problem to its solution, i.e. the existence of the so-called "universal problem solver". This too narrows down the sphere of its constructive applicability.

Brouwer showed, that in principle it is possible to develop mathematics shunning the non-constructive principles, in particular—the principle of omniscience and that of the universal problem solver. The idea of logic as a calculus of problems, and that of the logical rules as transformers of problems into solutions, was made more exact by A.N. Kolmogorov [7]. Of late it has been realized that an enormous advantage of Kolmogorov's interpretation lies in the fact that it has not been specified through to the end, and, thus, it is rather an outline that can be modified. But to begin with this advantage was taken for a deficiency; both S.C. Kleene and A.A. Markov derived the constructivist (intuitionist) logic from an exact definition of algorithm and of the concept of natural number. They showed that the principles against which Brouwer raised objections, are ineffective after a natural algorithm is given for singling out the constructive task from among arithmetical formulae, and an interpretation is provided for the transformations operative in this constructive task, as algorithms (partially—recursive functions). N.A. Shanin carried this construction through to its logical end [8] and, he provided an explicit algorithm for singling out the constructive task, having showed that after this substantiation of the construction carried out, one may go ahead with the methods of classical mathematics.

The third—significative—trend views mathematics as a science of formalisms and of the modes of their transformations. E. Schröder should be considered to be the spiritual father of this trend in many of its aspects, though he did not formulate an explicit "manifesto" of significative mathematics.

Now we shall dwell upon the paths of development, the mutual interactions and, the future perspectives of these three trends.

Gödel's incompleteness theorem turned out to be a turning point in the fate of the pseudo-classical trend. Naive hopes to the effect that it will be possible to get away without any serious reexamination of the paradigm of classical mathematics, having simply "sanctified" it with the Hilbertian incantations (and then one may happily forget about them)—were not vindicated. It became clear that the pseudo-classical trend too demands a serious reconstruction of the entire system of mathematical concepts, and that is why, from the point of view of psychological protection, the easiest thing to do was to simply interpret Gödel's [incompleteness] theorem as the collapse of Hilbert's programme as a whole—to be able to get along with what one was doing, this time, openly refusing to bother about the foundations. A quite frequent methodological error crops up in the interrelationships among the theoreticians and the "practitioners": the "practitioners" are inclined to demand that the theoreticians should substantiate their [i.e., practitioners'] positions and activities, but only a "theoretician" on the verge of becoming a charlatan can provide such substantiation, in so far as in practice there is always a mix up of the rational, extremely exact activities, not only with the non-optimal, superfluous moves, but also with the plainly bad ones, dictated by tradition. That is why a true theory cannot substantiate practice, but must reconstruct it.

Nevertheless, namely, in this slack period, principled results were obtained by P.S. Novikov [9]: he established that from the classical proof of a formula of the form $\exists\, x\, A\,(x)$ in arithmetic—where $A(x)$ is algorithmically solvable—it is possible to obtain the construction of an n, such that $A(n)$.

During the last few years, the demands of theoretical programming, informatics and those of the so-called "artificial intelligence" have forced the logicians to return to the pseudo-classical trend. In particular, the development of the programming languages like PROLOG [10], has put forward the task of isolating those sub-systems of the classical logic, which retain constructivisability in some sense—as one of the most important tasks in this field. It appeared, in particular, that the system of Horn's formulae of the form

$$\forall x_1, \ldots, x_n\, (P_1 \,\&\, \ldots \,\&\, P_k \Rightarrow Q), \tag{2}$$

upon which PROLOG is based, possesses such property.

The classical formulae, which provide on opportunity to describe and elicit constructivist constructions, have been more systematically described in the latest works of the Novosibirsk school: [11], [12]. Here we find the theoretical foundations for all possible systems of logical programming, based upon classical logic. In yet another work [13] semantic conditions have been introduced upon the constructivisation of classical theories, without imposing any limit upon the class of formulae used. It appeared that for the fully constructive models (where any set defined by a formula of our label (signature) is recursively solvable), the classical logic is completely constructively interpretable.

The development of constructivist logic has forced us to ear-mark yet another component in the definition of a calculus, besides the axioms and the rules of deduction: it is the global structure of the inferences; and this has opened up yet another opportunity for the development of the pseudo-classical trend. Even the intuitionist logic can be interpreted as classical logic with a limited global structure of natural deduction; it has been established [14] that one can obtain the classical logic from intuitionist logic by adding to it one global, structural rule of inference: the rule of accepted unexpectedness.

However, all these possibilities, opened up for the use of the pseudo-classical trend do not alter the basic conclusion of Brouwer: while using a logic one must carefully investigate the class of problems and the class of implied interpretations, otherwise reasoning and construction would inevitably drift apart. The ear-marked sub-classes of formulae, theories and inferences are based on such investigations in every case.

The pseudo-classical trend has unexpectedly turned up on the highway of development of classical mathematics itself. Though it has not been explicitly recognized, category theory may be that instrument, wherein again, as in the Hellenic mathematics, the constructive and the descriptive aspects have merged into one—where, proof guarantees construction. But this has now happened owing to a transition to a new level of abstraction, which has again permitted the banning of any explicit reference to the numbers. Further, the theory of categories leads to the necessity of investigating its own inner logic, which is intuitionst in the spaces (toposes) and coherent in the more general case [15]. Thus, here also, the pseudo-classical trend closes in on the non-classical trend.

Now, the non-classical trend has divided itself into two branches: intuitionism and constructivism.

In intuitionism we essentially base ourselves upon the incompleteness of our knowledge. Namely, we do not intend to provide any precise and final definition of the class of effective constructions [16]. What is more, in intuitionism we try to use this indeterminateness, this ignorance, as a positive factor. For example, from the substantiation of the principle of continuity in [16]:

$$\forall \alpha \exists n\, A(\alpha, n) \Rightarrow \forall \alpha \exists k\, \forall \beta\, (Vl\, (l < k \Rightarrow \alpha(l) = \beta(l))\, \&\, A(\beta, n) \Rightarrow A(\alpha, n)) \tag{3}$$

it is evident, that this principle signifies the absence of any knowledge of the global rules, which would indicate the behaviour of the Brouwerian "sequences of choice" or of the "sequences that have become free". What is more, later on a conception of "lawless sequences" has been worked out, wherein, in general, all accessible information constitutes an initial block [17]. Thus, in intuitionism an attempt is made to demonstrate that the knowledge of ignorance happens to be the most valuable form of knowledge. Intuitionism is thus sharply at variance with the entire paradigm of classical mathematics.

Constructivism tries to unite constructivisability with maximum retention of the classical mathematical paradigm. To some extent constructivism is as Platonist, as classical mathematics. The class of objects under consideration and the methods of their transformation (at this point we have an essential difference with classical mathematics—where one does not even think of the methods of transformation) are formulated precisely, basing the formulations upon an exact concept of algorithm. Knowledge is interpreted as a normal state, and ignorance—as an anomaly, which is inevitably present, but which must be overcome at all cost. Such an interpretation permitted A.A. Markov [18] to clearly ear-mark the system of initial abstractions, which are foundational to constructivist mathematics. This interpretation predetermined the journey of constructivist mathematics to a dead end in the narrow constructivism of N.A. Shanin [19], where an attempt has been made to totally ban ideal sentences from mathematics.

E. Bishop tried to occupy an intermediate position [20], when he tried to get away from an exact fixation of the class of effective methods, as well as from basing oneself upon ignorance. But when his conception was made more precise—see, in particular Per Martin-Löf's book [21]—it was found, that Bishop's conception lies completely within the frame-work of constructivism. Martin-Löf was the first to make use of the circumstance—though it is true that he did not formulate it explicitly—that the giving of the constructive objects and of an exact description of the methods of transformation of the objects, still does not fully determine the methods of transformation of methods, and that such constructive functionals of the higher type, metaalgorithms, can be varied completely—without touching the algorithms themselves.

During the last few years, the demands of application—in particular, of informatics,—has stimulated a Renaissance of Constructivism, even in our country; but this time it is a broad constructivism, which investigates the most diverse classes of methods and, correspondingly, the most divergent constructivist theories and even the constructivist logics. The very concept of constructivist logic is little by little tearing itself away from its unjustified ties with a single class of problems and effective methods and is becoming a relative concept, which may be varied, depending upon the descriptive language, in which transformable objects are described and problems are posed, and the class of programmes, or methods of transforming concrete and abstract objects are put forward. Viewed thus, intuitionist logic itself appears as the most classical of the constructivist logics—this logic is constructivist in that situation, where we are faced with the task of pure functional programming [22]. In other words, a constructivist use of intuitionist logic is possible, when we are not very much constrained in terms of time and other resources, when our computations only add new data, new knowledge, over and above what we already have and, when it is possible to use the complex structures of data and the functionals of higher type.

The use of independent constructivist logics, and not of some fragments of classical logic, is expedient, when thereby the expressive power of language is sharply raised, and when the constructivist description is by far shorter and clearer than the corresponding classical one. For example, in [23] a case of pure implicative logic has been considered, wherein the formulae have been constructed with the help of a single logical connective—implication—from amongst the prepositional letters, and a translation of the problem described into the traditional languages requires all the finite type of λ-terms, predicates upon them, and quantifiers.

Practically speaking, the third—significative—trend is also highly topical. Formal calculi and transformations of formalisms are widely investigated in mathematical linguistics and theoretical programming. But the methodological and metamathematical aspects of the significative point of view have not been sufficiently analysed . In this connection see the works of P. Lorenzen [24] and S. Yu. Maslov; [25], [26].

Let us attempt a few conclusions. It is possible that the arrival of the pseudo-classical, "Hellenic" stage—wherein the constructive and the descriptive aspects merge into one—is indicative of the maturity and conceptual unity, of the conceptual completeness, of the system of mathematical concepts. Here, explicit mention of numbers are banned from the theories.

The desire to "carry the results right upto the numbers", to have explicit theories of computational methods and, generally, of the methods of practical constructions, leads to constructivism—in one form or the other. The different problem-oriented constructivist theories must give rise to an unified pseudo-classical theory, describing the effective structures of a given class.

The intuitionist theories are more abstract and they too can give rise to an entire family of more concrete constructivist theories; but they are operative in a different situation: when it is not expedient to assume the completeness of a system of concepts or, when the system of concepts is explicitly non-formalizable and when there is no point in striving at completeness.

Finally, it is necessary to conduct serious and deep-going investigations, to ascertain the emerging shape of significative mathematics.

Literature

1. *Struik. D.J.*, A Concise History of Mathematics. N.Y., 1967 [Russ. Tr. M., 1984).
2. *Aristotle*, Works. Russ. cd., M., 1978. T. 2.
3. *Tarski A.*, A Decision Method for Elementary Algebra and Geometry. Berkley; L.A., 1951.
4. *Biryukov B.V., Biryukova L.G.*, "Uchenie o formakh (velichinakh)" Germana i Roberta Grassmanov kak predvoskhischenie knonstruktivnovo napravleniya v matematike. I. // Kibernetika i logicheskaya fomalizatsiya: Aspekty istorii i metodologii. M., 1982.
5. *Schröder F.*, Vorlesungen über die Algebra der Logik (exakte Logik). Leipzig, 1890-1905.
6. *Brouwer L.E.J.*, De onbetrouwbaarheid der logische principes. Bd. 2.
7. *Kolmogorov A.N.*, Zur Deutung der intuitionistischen Logik // *Math. Zeitschr.* 1932. Bd. 35.
8. *Shanin N.A.*, O konstruktivnom ponimanii matematicheskikh suzhdenii // Tr. MIAN SSSR, 1958. T. 52.

9. *Novikov P.S.*, On the Consistency of Certain Logical Calculus // *Mat. sb.* 1943. T. 12(54).
10. Sistema "PROLOG - ES". Kiev, 1979.
11. *Ershov Yu. L.*, Printisip Σ-perechisleniya // DAN SSSR. 1983. T. 270, No. 5.
12. *Goncharov S.S.*, *Sviridenko D.I.*, Σ-programmirovanie // Vychislitelnye sistemy. Novosibirsk. 1985. Vyp. 107.
13. *Voronkov A.A.*, Konstruktivnaya semantika dlya Teorii modelei // Vsesoyuz. Konf. po prikl. logike: Tez. dokl. Novosibirsk. 1985.
14. *Nepeyvoda N.N.*, Pravilo neozhindannosti i strukturnye *go to* // Semiotika i informatika. 1984. Vyp. 23.
15. *Goldblatt R.*, Toposy. Kategornyi analiz logiki. M., 1983.
16. *Heyting A.*, Intuitionism. Amsterdam. 1956. (Russ. Tr. M., 1965].
17. *Dragalin A.G.*, Matematicheskii intuitsionizm; Vvedenie v teoriyu dokazatelstv. M., 1979.
18. *Markov A.A.*, Teoriya algorifmov // Tr. MIAN SSSR. 1954. T. 42.
19. *Shanin N.A.*, Roil poniyatiya algoritma v semantike arifmeticheskikh yazykov // Algoritmy v sovremennoi matematike i eo prilozheniyakh. Novosibirsk. 1982. Ch. 2.
20. *Bishop F.*, Foundations of Constructive Analysis. N.Y., 1967.
21. *Martin-Löf P.*, Ocherki po konstruktivnoi matematike. M., 1975.
22. *Nepeyvoda N.N.*, Logicheskii podkhod k programmirovaniyu // Algoritmy v sovremennoi matematike i ee prilozheniyakh. Ch. 2. Novosibirsk, 1982.
23. *Nepeyvoda N.N.*, Konstruktivnye logiki // Neklassicheskie logiki. M., 1982.
24. *Lorenzen P.*, Einführung in die operative logik und Mathematik. B., 1954.
25. *Maslov S. Yu.*, Mutatsionnye ischisleniya: Zap. nauch. seminarov LOMI .L., 1975. T. 49.
26. *Maslov S. Yu.*, Deduktivnye sistemy i ikh ekonomicheskie primeneniya. M. Nauch. sovet po kompleks. probl. "Kibernetika" AN SSSR, 1983. Prepr.

Source: Zakonomernosti razvitiya sovremennoi matematiki. "Nauka". M., 1987. s. 219-229.

Author: Nikolai Nikolayevich Nepeyvoda (1949-), Soviet and Russian mathematician, logician, information scientist and author of science fictions.

Appendices: Two bibliographies

APPENDIX I
A BIBLIOGRAPHY ON HEGEL AND MATHEMATICS

PRIMARY LITERATURE

Abbreviation:

G W: G.W.F. Hegel, *Gesammelte Werke* [*Collected Works*] (1968-); Hamburg: Felix Meiner Verlag.

Hegel, Georg Wilhelm Friedrich (1936), "Geometrische Studien",in: *Dokumente zu Hegels Entwicklung* [*Documents on Hegel's Development*], Hrsg. Johannes Hoffmeister: 288-300; Stuttgart: Fr. Fromanns Verlag.

English translation:

__ (2008), "Geometrical Studies", translated with introduction and notes by Alan Leonard Tuke Paterson, *Hegel Bulletin,* Volume 29, Issue 1-2 (Number 57/58) 2008: 132-153.

__ (2014), Geometrische Studien: Zwei Fragmente [Geometrical Studies: Two Fragments], in: *G W,* Erste Abteilung, Band 2, Frühe Schriften [Early Writings], Teil II: 369-384.

__ (1978), *Wissenschaft der Logik* [*Science of Logic*], Erster Band, Die Objektive Logik [The Objective Logic] (1812/1813), *GW,* Erste Abteilung, Band 11.

__ (1981), *Wissenschaft der Logik* [*Science of Logic*], Zweiter Band, Die Subjektive Logik [The Subjektive Logic] (1816), *GW,* Erste Abteilung, Band 12.

__ (1984), *Wissenschaft der Logik* [*Science of Logic*], Erster Teil, Die Objektive Logik, Erster Band, Die Lehre vom Sein [First Part, The Objective Logic, First Volume, The Doctrine of Being] (1832), *GW,* Erste Abteilung, Band 21.

English translation:

__ (2010), *The Science of Logic,* translated and edited by George Di Giovanni; Cambridge etc.: Cambridge University Press.

__ (2001), *Enzyklopädie der Philosophischen Wissenschaften im Grundrisse* (1817), *GW,* Erste Abteilung, Band 13.

Partial English translation:

__ (2010), *Encyclopedia of the Philosophical Sciences in Basic Outline,* Part I: Science of Logic, translated and edited by Klaus Brinkmann and Daniel O. Dahlstrom; Cambridge etc.: Cambridge University Press.

__ (2013), *Vorlesungen über die Wissenschaft der Logik* [*Lectures on the Science of Logic*] I, *GW,* Zweite Abteilung, Band 23, 1.

__ (2015), *Vorlesungen über die Wissenschaft der Logik* [*Lectures on the Science of Logic*] II, *GW,* Zweite Abteilung, Band 23, 2.

__ (2017), *Vorlesungen über die Wissenschaft der Logik* [*Lectures on the Science of Logic*] III, *GW,* Zweite Abteilung, Band 23, 3.

__ (2017), *Katalog der Bibliothek Georg Wilhelm Friedrich Hegels* II, Abteilung V. Mathematik, Naturwissenschaften und Medicin [*Catalog of the Library of Georg Wilhelm Friedrich Hegel* II, Section V. Mathematics, Natural Sciences and Medicine]: 1355-1602, in: *GW,* Zweite Abteilung, Band 31, 2.

A link to a web page on
George Wilhelm Friedrich Hegel, *Science of Logic*:
<https://ncatlab.org/nlab/show/Science+of+Logic>; at:
<https://ncatlab.org/nlab/show/HomePage>
contains links to a lot of primary and secondary literature.

SECONDARY LITERATURE

Alcouffe, Alain (1985), "Marx, Hegel, et le <<calcul>> ["Marx, Hegel, and the <<calculus>>"], in: *Les Manuscrits Mathématiques de Marx* [*The mathematical manuscripts of Marx*], pp. 9-108; Paris: Economica.

Baer, Reinhold (1932), "Hegel und die Mathematik" ["Hegel and Mathematics"], S. 104-120, in: *Verhandlungen des zweiten Hegelkongresses vom 18. bis 21. Oktober1931 in Berlin* [*Proceedings of the second Hegel congress from 18 to 21 October 1931 in Berlin*], Hrsg. Baltus Wigersma; Tübingen: J.C.B. Mohr etc. URL: <http://hegel.net/andere/baer_hegel_math.pdf>

Beach, Edward (2006), "Hegel's Misunderstood Treatment of Gauss in the Science of Logic", *Idealistic Studies,* 36 (3): 191-218.

Carlson, David Gray (2007), A Commentary to Hegel's *Science of Logic,* New York: Palgrave Macmillan.

Damsma, Dirk (2008), On the dialectical foundations of mathematics, Version 5.3; Amsterdam: Universiteit van Amsterdam. URL: <https://pure.uva.nl/ws/files/1314370/96369_286fulltext.pdf>

__ (2011), Version 5.2 of Damsma (2008); URL: <https://pure.uva.nl/ws/files/1392167/113450_355278.pdf>

__ (2009), Marx's systematic dialectics and mathematics and their articulation in his 'schemes of reproduction'; Amsterdam: Universiteit van Amsterdam. URL: <https://pure.uva.nl/ws/files/1277549/98763_1072fulltext.pdf>

__ (2010), Qualitative and Quantitative Analysis in Systematic Dialectics: Marx vs. Hegel and Arthur vs. Smith; Amsterdam: Universiteit van Amsterdam. URL: <https://pure.uva.nl/ws/files/1042649/97895_1071fulltext.pdf>

__ (2011), "Set Theory and Geometry in Hegel", *Hegel-Jahrbuch,* 2011, Issue 1: 54-58.

Frantz, Constantin (1842), *Die Philosophie der Mathematik. Zugleich ein Beitrag zur Logik und Naturphilosophie* [*The Philosophy of Mathematics. At the Same Time a Contribution to Logic and Natural Philosophy*]; Leipzig: Hermann Hartung.

Kolman, E. and S. Yanovskaya (1931), "Gegel' i matematika" ["Hegel and Mathematics"], *Pod znamenem marksizma*, No. 11-12 (Noyabr'-Dekabr' 1931): 107-120.

Lacroix, Alain (2000), "The Mathematical Infinite in Hegel", *Philosophical Forum,* 31 (3/4): 298-327.

Lawvere, Francis William (1991), "Some thoughts on the future of category theory", in: *Category Theory*; Berlin etc.: Springer Verlag.

__ (1996a), "Unity and Identity of Opposites in Calculus and Physics", *Applied Categorical Structures,* 4: 167-174.

__ (1996b), "Grassmann's Dialectics and Category Theory", in: Gert Schubring [Ed.], *Hermann Günther Graßmann (1809-1877): Visionary Mathematician, Scientist and Neohumanist Scholar,* 255-264; Dordrecht: Springer.

__ Homepage of Francis William Lawvere: <http://www.acsu.buffalo.edu/~wlawvere/index.html>

Lawvere, Francis William and Stephen Hoel Schanuel (1997), *Conceptual Mathematics: A first introduction to categories*; Cambridge: Cambridge University Press.

Paterson, Alan Leonard Tuke (1997a), "Towards a Hegelian Philosophy of Mathematics", *Idealistic Studies,* 27 (1/2), 1997: 1-10.

__ (1997b), "Self-reference and the natural numbers as the logic of Dasein", *Hegel-Studien,* 32 (1997): 93-121.

__ (2000), "The Successor Function and Induction Principle in a Hegelian Philosophy of Number", *Idealistic Studies,* 30 (1): 25-60.

__ (2002), "Does Hegel Have Anything to Say to Modern Mathematical Philosophy?", *Idealistic Studies,* 32 (2): 143-158.

__ (2004/2005), "Hegel's Early Geometry", *Hegel Studien,* 39/40: 61-124.

Drafts of the following texts are available on the webpage of Alan Leonard Tuke Paterson:<https://sites.google.com/site/apat1erson/>:

__ (No Date), Frege and Hegel on concepts and number;

__ (2007), The Hegelian Concept and set theory;

__ (2014), The Hegelian logic implicit in formal logic; and,

__ (June 2018), Introduction to *The philosophy of self-reflexivity.*

Petry, Michael John [Ed.] (1993), *Hegel and Newtonianism,* Part Two: Mathematics: 139-225; and Part Seven: Bibliographical: 669-719 (especially, the pages 670-687 and 711-716 pertaining to mathematics therein); Dordrecht: Springer.

Rodin, Andrei (2014), *Axiomatic Method and Category Theory*; Cham etc.: Springer International.

Szigeti, Jozsef (1972), Hegel und G. Cantor [Hegel and G. Cantor], in: *Hegel-Jahrbuch1971*: 283-293; Meisenheim am Glan: Anton Hain Verlag.

APPENDIX II
A BIBLIOGRAPHY ON MATHEMATICS AND SEMIOTICS

Alexander, James C. (2011), "Blending in Mathematics", *Semiotica,* Volume 2011, Issue 187: 1-48.

Duval, Raymond (2017), *Understanding the Mathematical Way of Thinking – The Registers of Semiotic Representations*; Cham: Springer.

Hausberger, Thomas (2015), "Abstract Algebra, Mathematical Structuralism and Semiotics", in: Krainer, Konrad and Naďa Vondrová [Eds.], *Proceedings of the Ninth Congress of the European Society for Research in Mathematics Education* (CERME 9, 4-8 February 2015): 2145-2151; Prague, Czech Republic: Charles University, Faculty of Education and ERME.

Ivanov, Vyacheslav Vsevolodovich (2007), "Towards Semiotics of Number", *Bulletin of the Georgian National Academy of Sciences*, Volume 175, No. 1 (2007): 186-194.

Lakoff, George P. and Rafael E. Núñez (2000), *Where Mathematics Comes From*; New York: Basic Books.

Manin, Yuri Ivanovich (2007), *Mathematics as metaphor: selected essays of Yuri I. Manin*; Providence, Rhode Island: American Mathematical Society.

Presmeg, Norma, Luis Radford, Wolff-Michael Roth and Gert Kadunz (2016), *Semiotics in Mathematics Education*; Cham: Springer Open.

__ [Eds.] (2018), *Signs of Signification: Semiotics in Mathematics Education Research*; Cham: Springer.

Radford, Luis, Gert Schubring and Falk Seeger [Eds.] (2008), *Semiotics in Mathematics Education: Epistemology, History, Classroom, and Culture*; Rotterdam/Taipei: Sense Publishers.

Roth, Wolff-Michael and Luis Radford (2011), *A Cultural-Historical Perspective on Mathematics Teaching and Learning*; Rotterdam/Taipei: Sense Publishers.

Rotman, Brian (1980), *Mathematics as Language: An essay in Semiotics* [Unpublished].

__ (1987), *Signifying Nothing: The Semiotics of Zero*; London: Macmillan Press.

__ (2000), *Mathematics as sign: writing, imagining, counting*; Stanford, California: Stanford University Press.

Sánez-Ludlow, Adalira and Gert Kadunz [Eds.] (2016), *Semiotics as a Tool for Learning Mathematics: How to Describe the Construction, Visualisation, and Communication of Mathematical Concepts*; Rotterdam/Taipei: Sense Publishers.

Subject Index

Name Index

[This index includes the names of some of the texts, journals and institutions indicated in the volume.]